PARABOLA

THE MAGAZINE OF MYTH AND TRADITION

VOLUME XVI, NUMBER 1,
FEBRUARY, 1991

Y0-ARK-139

MONEY

Founder and Editorial Director
D. M. DOOLING
Co-Editors
ROB BAKER, ELLEN DRAPER
Executive Publisher
JOSEPH KULIN

Senior Editors
LEE B. EWING, PAUL JORDAN-SMITH
JEAN SULZBERGER, PHILIP ZALESKI
Contributing Editors
MARVIN BARRETT, JOSEPH EPES BROWN
JOSEPH BRUCHAC, JONATHAN COTT
FREDERICK FRANCK, WINIFRED LAMBRECHT
DAVID LEEMING, LOBSANG LHALUNGPA
RICHARD LEWIS, ROGER LIPSEY
JOHN LOUDON, BARRE TOELKEN
P.L. TRAVERS, JEFFREY P. ZALESKI

Manager, Advertising and Mailing Lists
BETH LEONARD

Manager, Fulfillment and Customer Relations
PETER FRANDSEN
Fulfillment Assistant
STEPHANIE PIASECZYNSKI
Senior Administrative Assistant
STANLEY EDWARDS

Art Director
DANIEL J. MCCLAIN
Mechanical Artists
JERRY TEEL, JOAN CARNEY
Art Research/Editorial Assistant
ROBERT YAGLEY
Proofreader
THOMAS V. FLATT

PARABOLA (ISSN: 0362-1596) is published quarterly by the Society for the Study of Myth and Tradition, Inc., a not-for-profit organization. Contributions are tax-deductible.

Subscription Rates: Single issue: $6.00. By subscription: $22.00 yearly, $38.00 for two years, $54.00 for three years. Postage for outside territorial U.S.: add $6.00 for surface rates, $20.00 for air, per year. **Postmaster:** Send address changes to PARABOLA, 656 Broadway, New York, NY 10012-2317. Second class postage paid at New York, N.Y., and additional offices.

Address all correspondence regarding subscriptions and advertising to PARABOLA, 656 Broadway, New York, NY 10012; Tel: (212) 505-6200; Fax: (212) 979-7325. Correspondence regarding editorial matters ONLY should be sent to The Editors, PARABOLA, 369 Ashford Ave., Dobbs Ferry, NY 10522; Tel: (914) 693-7960; Fax: (914) 693-7978. PARABOLA assumes no responsibility for unsolicited material. Manuscripts or artwork not accompanied by stamped, self-addressed envelopes will not be returned.

Distributed in the United States and Canada by Eastern News, 1130 Cleveland Road, Sandusky, Ohio 44870. Other foreign distribution is handled by PIMS, 105 Madison Avenue, New York, NY 10016.

Microform copies of PARABOLA may be obtained from University Microfilms, Inc. PARABOLA is indexed in the *Religion Index of Periodicals, Abstracts of English Studies, Book Review Index, Current Contents, American Humanities Index,* and *CERDIC.*

VOLUME XVI, No. 1, FEBRUARY 1991

COVER: *"The Moneychanger and his Wife"* by Quinten Metsys, early sixteenth-century Flemish painter, at the Musée du Louvre, Paris.

FOCUS

T HE IMMEDIATE associations with the word "money" are emotional and usually negative: greed, indebtedness, bankruptcy, "filthy lucre," the love of it being the "root of all evil" . . . Yet, as Sri Aurobindo states, money, "in its origin and true action belongs to the Divine." In pondering the theme for this issue of PARABOLA, we have been struck by just such negative and positive contrasts. This has brought the question about what money *is,* and what it could be in its "origin and true action."

In this process of examination, the very terms we employ to try to speak about money begin to take on a new level of meaning. What is *interest*—where does it come from? How do we *value* something? What do we do when we *invest*? What takes place in an *accounting*? What are our *debts,* which perhaps can be forgiven through right action? Where is *credit* due in our lives? How is something *charged*? What is *currency*? A *gratuity*? *Payment*? *Home economics*? What do we really confront when "the buck stops here"?

Money is a basic symbol in our lives: it represents energy, power, and the potential for real action. Its original uses seem to have included a rec-

ognition of true value, and a requirement for payment as a relating factor between people, and between people and the Divine. In this exchange, a kind of circulation was established, like that of the blood in the human body. Thomas Buckley speaks of such free movement of energy in describing the symbols of exchange among the Yurok Indians of Northern California, where everything of value becomes a part of the dance of life and wealth itself "cries to dance."

A few rare souls, like Mother Teresa, have managed to remain free of the attachment to money, but most people's lives have been taken over by it; they have become prisoners to their own emotional obsession with a symbol that has lost its deeper and higher meaning. Very few can resist "the old ninety-nine rupee snare," having *almost* "enough" so that one is just on the verge of satisfaction, but never satisfied. On the other hand, in the Indian tale "The Simple Grasscutter," a peasant farmer manages to teach two wealthy rulers the true meaning of value; when he is rewarded with a castle of his own, he goes back to the fields where he smells "the sweetest grass."

Real value is elusive but essential, like "The Spirit in the Bottle," a Grimm's fairy tale which C.G. Jung has suggested symbolizes alchemical transformation. Long hard work is met by the release of a higher energy, resulting in a metaphor of money's power to do: an ordinary axe turned into pure silver.

In today's topsy-turvy world, how can value and payment be rediscovered, how can we "redeem our relationship" to money, as Joseph Kulin suggests? The story "The Miser" indicates that such a shift in attitude is possible, but only as the result of suffering and struggle, the undertaking of an arduous process. This process is the movement from attachment to freedom, as described in Karl Potter's article on the Hindu attitudes of *artha, kama, dharma,* and *mokṣa.*

Money demands a struggle in order to be rightly valued, but it can also be a reminder, a spur of conscience, like P.L. Travers' threepenny bit. And we can be reminded of its deeper meaning both by the hustle and bustle of the city, as Socrates insists in David Appelbaum's article, and by Nature in a state of balance, as Native American traditions acknowledge.

As agent of enslavement or as invitation to liberation, the double-edged sword of the theme never loses its keenness. The jingle of coins evokes both our basest attitudes and our highest aspirations, the alchemical lead and gold present in all of us, and which must be there in order for transformation to take place.

—*Rob Baker and Ellen Draper*

PARABOLA
SPRING 1991

The One Who Flies All Around the World

Thomas Buckley

THE NATIVE PEOPLES of northwestern California (I will be writing mostly about the Yurok Indians) had "money" long before non-native invaders, lusting for gold, brought them nickels, dimes and dollars—and many new ways to die. Before minted United States currency was introduced in the 1850s, the primary monetary medium in the area was dentalia shells—the hard, phallic, tusk-like shells of *Dentalium indianorum* and *D. hexagonum* that were traded slowly down from Vancouver Island and Puget Sound. The shells gained in value the farther they traveled, distance and scarcity providing reliable hedges against inflation and devaluation until the introduction of coastwise shipping, after the gold rush.

Five sizes of shells were recognized as money (*chi:k*) by the Yuroks. The largest and most valuable were two and a half inches long, or eleven to a twenty-seven-and-a-half-inch string. The next, twelve to a string, were worth about one fifth as much, and so on down to the smallest, fifteen to a string. Shells smaller than that were strung as beads in necklaces but were not used by men in exchange (money, we will see, was largely men's business). The longest and best shells were often decorated with incised lines or with bits of snakeskin and red woodpecker scalp.

Woodpecker scalps themselves were another form of money, called *chi:sh*. The large scalps of pileated woodpeckers (*Dryocopus pileatus*) were the most valuable, but the scalps of the much smaller acorn woodpecker (*Melanerpes formicivorus*) also had an established exchange value.

All of this money did not comprise a truly "generalized medium of exchange," as anthropologists define "money," but was part of a far more complex system of wealth. As far as money itself goes, there was men's money and, of lesser general import, women's money, as well as other things, like food, that were themselves forms of wealth. Money from one system was not used to pay for things from another system. Women's money might buy baskets, for instance, which were made by women, and only the most desperate and *déclassé* sold food for shells or feathers. Men's

money was quintessentially used to pay for human life: for births, through bridewealth, and for deaths, through blood money. Monetary systems coexisted with less formal systems of barter and gifting. Thus men could at once be extraordinarily single-minded and astute in the pursuit of certain forms of wealth, yet hold greed and stinginess to be the moral failings of no-accounts.

Despite profound differences between their monetary economies, Indian people in northwestern California understood quite readily what "white man money" was when they were introduced to it after contact with whites, at "the end of Indian time." Minted coins entered the indigenous economy almost immediately, after 1850, and inter-monetary equivalents were soon established. In the 1860s a dentalium shell of the largest size was worth five American dollars and a full string of them about fifty dollars. The next-to-largest shells went for one dollar apiece, the same as a pileated woodpecker scalp (the Karuks called these woodpeckers "dollar birds" well into this century). A wife of good repute (or, really, the children that she would bear) cost about three hundred dollars; more was required to pay for the killing of a man of substance.

Men once kept their accumulated dentalia in elkhorn purses that were secretly compared to a woman's uterus. They sought to accumulate as much as possible through asceticism and, more directly, through shrewd manipulation of the local legal system, through the marriages of close female kin, and through gambling. When white man money came, the same practices continued, Indians now using it as a medium. So, the people who followed the old ways may seem much like the "primitive capitalists" that earlier anthropologists and psychologists portrayed them to be. But wait. The Yurok mythic creator Great Money, *pelin chi:k*, both helped to make the world and had the capacity to destroy it. To understand these supposed "primitive capitalists" we must leave off the study of economics *per se* and take up mythology.

GREAT MONEY, a gigantic winged dentalium, is one of the creators of the world. He comes from the North and is the oldest and biggest of five brothers (the five sizes of dentalium money). Together with the next oldest and largest, *tego'o*, Great Money once traveled and traveled, directing the work of the other creators. His wife is Great Abalone, women's money from the South.

Great Money is perhaps the most powerful of all the creators. He alone can fly around the entire world, seeing all of it and everything that happens here, and he has the frightening ability to swallow the entire cosmos should he choose to. (Even the smallest Money Brother can do this if offended badly enough.) Great Money has beautiful wings which he is loath to soil, and when he travels by boat he sits in the middle doing nothing, like a very rich man who deigns neither to paddle nor to steer. The other creators did all the actual work in the beginning, under Great Money's direction. One of them, the trickster Wohpekumew, gave money itself to humankind. He said, "A woman will be worth money. If a man wants to marry her, he will pay with money. And if a man is killed, they will receive money from [the killer] because he who killed . . . must settle" (A.L. Kroeber, *Yurok Myths*, 1976). Again, money was brought into the world that Great Money helped to create to pay for births and deaths, for human life, for Life itself.

In the old days, I think, money in northwestern California was a many-layered symbol for what we might call "life-force" or "spirit." This "spirit" (*wewolochek'*) was understood as a male contribution to human existence. When one sought legitimate children one paid for their lives with bridewealth, balancing an account with the mother's natal family which had, in a sense, "lost" her life-force. (Paid-for children "belonged" to the father's family, and unpaid-for children, "bastards," had few rights, were viewed as barely *alive*.) By the same token, one could kill an enemy for one's own satisfaction if one could pay his family for his life, through blood money, replacing the life one had taken with what was, I am saying, symbolically more of the same. One who could not pay his debts became a slave, had no life of his own. "Everything has a price," said an old man I knew.

To BETTER understand money-as-life, or as spirit, we need to consider the broader category of "wealth," for there is far more to being "rich," among these people, than merely having a lot of money.

What the Yuroks and their neighbors have always valued is a balanced variety of wealth, not simply an abundance of money. For example, while in theory a wife might be worth three hundred American dollars, such a simple payment would be without honor. Rather, several strings of dentalia, some good dance regalia, a well-made boat, as well as some gold coins (after whites brought them), with a total value of about three hundred dollars, made a prestigious payment and assured the high standing of the children that followed. The death of a high status man might cost the killer fifteen strings of dentalia, but also valuable regalia and, perhaps, a daughter.

Money, then, dentalia and woodpecker scalps, was once only the beginning of a continuum that included and perhaps culminated in ceremonial dance regalia. These regalia used, today as in the past, in the child-curing Brush Dance and the world-fixing Jump and Deerskin Dances, include dentalia necklaces and headrolls ("headdress" is not a term used by local people, who say "we don't wear dresses on our heads") richly decorated with woodpecker scalps. There are fifty-three large woodpecker scalps on a Jump Dance headroll. But these regalia also include beautiful feathers of many other kinds, rare hides and pelts (especially the priceless hides of white deer), large, finely wrought obsidian points, special Jump Dance baskets shaped like purses, with labia, and much else.

Women own beautifully made buckskin dance dresses worked with haliotis, olivella, and clam shell, pine nuts and juniper berries, with braided beargrass tassles, worn together with loop after loop of shell necklaces. As the girls and women walk and dance, the dresses sing expressive and unique percussive songs. All of this is wealth, continuous with money: beginning in the nineteenth century people used whole and halved American coins in decorating both men's and women's regalia, using the pierced metal pieces in the places where dentalia alone had formerly been used.

Non-locals tend to talk about ostentatious ceremonial "display" of wealth by the Yuroks and their neighbors. But Indians speak of the things themselves dancing: "This headroll danced last year at Hoopa"; "These feathers want to dance, so I'm going to let them go to Klamath." *The regalia is alive and feeling.* Traditionally, people get money, feathers, and other dance things by "crying" for them—practicing austerities and seeking the compassion of the Immortals, who send wealth to the righteous. And the wealth itself, once in hand, "cries to dance." People say, "Those feathers were crying to dance, so I let them out to go to Katimin," to the Brush Dance there.

Finally, while people cry for wealth, and the wealth cries to dance, the Immortals themselves "cry to see" the beautiful things dance. "Give them what they cry for!" said a Hupa acquaintance at a dance, not long ago. The intricate symmetry of interdependencies bespoken in all of this is delightful and deep.

The Immortals join the human beings, invisibly, to watch the dances, standing in special places reserved for them alone, happy to see the "stuff" dance. The big dances, solemn and powerful, "fix the world," restore its harmony, ease suffering, fend off disease and hunger, assure more life. Anthropologists have called these dances "world renewal dances." To read their symbolism in the most direct way, we might say that the dancing of accumulated "life" or "spirit," emblematized by wealth, affirms existence; it restores the balance of life in the cosmos, reversing the nearly inexorable tendency of the world toward entropy and death by paying off the deficit accumulated through innumerable breaches of sacred law—settling the energy account again, somewhat as deaths are paid for. This is true, too, in the Brush Dances, where the dancing of regalia helps restore a sick child to life. "Give them what they cry for!" Pay back to "The Spiritual" (some people call it) that upon which we depend, so that we may continue to exist. It makes sense, then, that some people assert that the Jump Dance, one of the "world renewal" dances, is also a "fertility dance," praying for and celebrating, especially, male procreativity. It is through men that the "life" of the cosmos enters the wombs of women, in the local view.

Before the white invasion, people converted money into wealth, stringing surplus small dentalia into necklaces, sewing woodpecker (and hummingbird!) scalps onto headrolls, or using such money to pay others to make fine regalia—wealth—for them. After the invasion and well into this century, however, the world seemed beyond fixing to many Indian people. Impoverished by the new economics of scarcity inaugurated by the whites, discouraged by massive losses of population and territory and power, many people converted their ritual wealth back into money, selling off their regalia, cheap, for white man dollars, or pawning it to those with better luck. Regalia left the world, stopped dancing.

Today these transformations con-

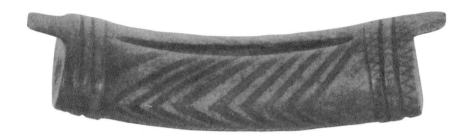

Elk-antler purse for dentalium money, Yurok Indian.

tinue. With slowly improving fortunes in the region, people are once again converting their money—now federal currency—into dance regalia, commissioning a new generation of skilled creative regalia makers and artists to make new wealth and gaining fame, as in the old days, by letting it dance. Brush, Jump and Deerskin Dances are once more the lively focus of Indian ways, and the esteem in which "dance makers"—those who own a large amount of regalia and have the connections to borrow even more—are held is high indeed.

Now, as probably always, money and wealth alike mean many things to different people, and those many meanings continue changing through time, in history, where these and all people are located. But I have risked a generalization.

Dentalium shells, men's money (compared sometimes to the male member), woodpecker scalps of the brightest scarlet, unmistakably like the blood of life itself, include among their many meanings mortal life-force, the energy of reproduction and continuation. From an Indian point of view, I believe, this "money," not love, makes the world go 'round.

There is a great, cosmic account that must be kept in balance if the people are to continue to survive. (The Karuk artist Brian Tripp entitled a recent work "The Best Things in Life Aren't Free.") In bridewealth, death settlements, and dancing, money and other forms of wealth are paid back into that great account. "Everything has a price" does not mean that everything is for sale, but that there is a necessary energic balance that must be maintained, because it is what makes mortal life possible and allows it to run through the generations.

The Indian peoples of northwestern California—Yuroks, Hupas, Karuks, Tolowas, and others—have tended to use money and wealth alike as master symbols of that which creates and enlivens the great diversity of meaningful forms in the world. Theirs has been, if you will, a spiritual economy. Non-Indian Americans, I think, have generally and ever-increasingly tended to reduce the many meanings and values of life-forms to the arbitrary monetary exchange values of those forms, putting virtually everything up for sale: a radically material economy. There is an enormous difference. ●

The Role of Money

Sri Aurobindo

MONEY IS the visible sign of a universal force, and this force in its manifestation on earth works on the vital and physical planes and is indispensable to the fullness of the outer life. In its origin and its true action it belongs to the Divine. But like other powers of the Divine it is delegated here and, in the ignorance of the lower Nature, can be usurped for the uses of the ego or held by Asuric[1] influences and perverted to their purpose. This is indeed one of the three forces—power, wealth, sex—that have the strongest attraction for the human ego and the Asura and are most generally misheld and misused by those who retain them. The seekers or keepers of wealth are more often possessed rather than its possessors; few escape entirely a certain distorting influence stamped on it by its long seizure and perversion by the Asura. For this reason most spiritual disciplines insist on a complete self-control, detachment and renunciation of all bondage to wealth and of all personal and egoistic desire for its possession. Some even put a ban on money and riches and proclaim poverty and bareness of life as the only spiritual condition. But this is an error; it leaves the power in the hands of the hostile forces. To reconquer it for the Divine to whom it belongs and use it divinely for the divine life is the supramental way for the Sadhaka.[2]

You must neither turn with an

ascetic shrinking from the money power, the means it gives and the objects it brings, nor cherish a rajasic[3] attachment to them or a spirit of enslaving self-indulgence in their gratifications. Regard wealth simply as a power to be won back for the Mother[4] and placed at her service.

All wealth belongs to the Divine and those who hold it are trustees, not possessors. It is with them today, tomorrow it may be elsewhere. All depends on the way they discharge their trust while it is with them, in what spirit, with what consciousness in their use of it, to what purpose.

IN YOUR personal use of money, look on all you have or get or bring as the Mother's. Make no demand but accept what you receive from her and use it for the purposes for which it is given to you. Be entirely selfless, entirely scrupulous, exact, careful in detail, a good trustee; always consider that it is her possessions and not your own that you are handling. On the other hand, what you receive for her, lay religiously before her; turn nothing to your own or anybody else's purpose.

Do not look up to men because of their riches or allow yourself to be impressed by the show, the power or the influence. When you ask for the Mother, you must feel that it is she who is demanding through you a very little of what belongs to her and the man from whom you ask will be judged by his response.

If you are free from the money-taint but without any ascetic withdrawal, you will have a greater power to command the money for the divine work. Equality of mind, absence of demand and the full dedication of all you possess and receive and all your power of acquisition to the Divine Shakti[5] and her work are the signs of this freedom. Any perturbation of mind with regard to money and its use, any claim, any grudging is a sure index of some imperfection or bondage.

The ideal Sadhaka in this kind is one who if required to live poorly can so live and no sense of want will affect him or interfere with the full inner play of the divine consciousness, and if he is required to live richly, can so live and never for a moment fall into desire or attachment to his wealth or to the things that he uses or servitude to self-indulgence or a weak bondage to the habits that the possession of riches creates. The divine Will is all for him and the divine Ananda.[6]

In the supramental creation, the money-force has to be restored to the Divine Power and used for a true and beautiful and harmonious equipment and ordering of a new divinized vital and physical existence, in whatever way the Divine Mother herself decides in her creative vision. But first it must be conquered back for her, and those will be strongest for the conquest who are in this part of their nature strong, and large, and free from ego; and surrendered without any claim or withholding or hesitation, pure and powerful channels for the Supreme Puissance.

●

NOTES

1. *Asura* designates the demon or anti-god.
2. Seeker.
3. Rajah-like.
4. The Divine Power.
5. Another name for the Divine Power.
6. Grace.

Reproduced from Sri Aurobindo, *The Mother* © Sri Aurobindo Ashram Trust (Pondicherry, India: Sri Aurobindo Ashram, 1974). Reprinted with the kind permission of Sri Aurobindo Ashram Trust.

EPICYCLE

TOM WHITE

The Miser/Jewish

THERE WAS ONCE a man who by hard striving had raised himself and his family from early poverty to fairly comfortable circumstances. He was now established as a *M'sader K'dushin*—a person qualified to perform the marriage ceremony. Because of his fine appearance and genial, graceful manner his services were preferred in many communities through the countryside.

His success had not freed him, however, from the dread of penury. He kept a strict hand on household expenditures, gave to charity only so much as might spare him criticism, and dressed sufficiently well from motives of shrewdness rather than self-respect or vanity.

When, in middle life, he inherited a good sum of money, which should have freed him from present stinting and from anxiety for the future, it actually had the opposite effect. He made of his strongbox a barrier against misfortune—a haven of safety. From adding to it as much as he could, and counting it up with satisfaction in the growth of the sum, he gradually came to take delight in fingering the coins themselves. He arranged them in piles—gold, silver, and copper—stacked them according to value and the year of minting, and as time passed, became unwilling to withdraw even small amounts. To meet expenses by taking money from the coffer caused him actual suffering.

He postponed giving to the community charities until he forgot his default. Only in cases of pitiful need that suddenly confronted him, giving him no time to consider, was he startled into generosity. Yet he knew himself for a religious person, and thought himself charitable because whenever a couple found it difficult to pay his fee, he would marry them without charge.

Small meannesses crept into his behavior. He would economize by fasting before he went to perform a ceremony, and gorging himself at the wedding feast. Nothing pleased him more than to fill his stomach at another's expense.

His gentle wife was saddened by this growing avarice. At one time, in the busy market, as she tried to stretch the sum he had given her for Sabbath preparations, she overheard some neighbors referring to him. They called him, not Hirschel "Glückenheimer," after the village of his birth, but Hirschel "*Kamzan*"—the "Miser."

She prayed that God would reveal to him the snare into which he had fallen. For she knew that her husband thought of his parsimony as "provision for the future" and "prudent care for his wife and children in later years." She had not the heart to reproach him, nor the words in which to point out his self-deception.

O NE DAY a handsome coach-and-four drove up to the house and the personage who entered said to the M'sader K'dushin: "I have come myself to engage your services. I am to be host at a wedding party on my estate, which is almost a day's journey from here. But that need not inconvenience you. If you will be good enough to come with me, you shall remain with us overnight, and I will see to it that you return home in good time. Furthermore, you may name your own fee."

The stranger's attire and the style of his equipage suggested great wealth. Delighted at the prospect, Hirsch Glückenheimer soon joined his patron in the smart coach, and the gray horses tossed their heads and took the road out of town.

For several hours they drove along roads, then they mounted to hilly country. The carriage swung and jerked uncontrollably, and despite the well-cushioned seats, the M'sader K'dushin found it a wearisome ride. Darkness fell and the outlook became wilder. They seemed to be plunging through an upland forest. A full moon rose and was visible at moments between tangled branches. But it brought no comfort. It looked down with what seemed to be a grimace.

At length they came out upon a high, meadowlike heath, and Glückenheimer was greatly relieved to see, incongruous amid this wilderness but reassuring, a fine, solid mansion streaming with light and the sound of festivities.

When the host introduced him, the M'sader K'dushin realized that never before had he seen so prepossessing a company. Richly dressed in harmonious colors and perfect taste, they danced with grace and elegance. They conversed in mellifluous tones that blended with the music. They looked at him with gracious kindness and made him welcome.

He was presented to the bride, and for a moment they were left together. She came closer to him and said: "All present here are spirits. I alone am a mortal like yourself. As you value your soul, accept no gift or fee. Do not eat or drink in this house!"

A moment later she was smiling as the bridegroom took her hand and led her to join a minuet.

Glückenheimer found a seat and sank into it. He felt a chill on his skin, as if the air had turned to ice. Looking about him, he understood the charm of that first impression. Here was no clumsiness, no obesity, no harshness, no screeching laughter, no fumbling, feeble age, as would surely figure at any mortal wedding. Their perfection threatened him. He felt himself weak and in their power.

But he was not alone for long. The host sought him out and insisted that he take some refreshment after his tiring journey.

Sternly—and warily—the M'sader K'dushin refused. When they pressed him, he excused himself with the plea that it was his custom to fast for twenty-four hours before the ceremony, so that his prayers for the future happiness of the young couple might be more acceptable.

Everybody praised his wholeheartedness; and the host, seeing that he would not be persuaded to break his fast, suggested he might wish to retire.

Glückenheimer gratefully assented. He was brought to a luxurious chamber, where he lay down on a cloud-soft bed and drew the curtains around him. It was a strange night. He slept, but woke frequently, always with the feeling that some elusive presence was slipping beyond his field of vision, or that he had failed to hear a message of importance, or that a song had ceased the very moment of his awakening.

IN THE MORNING his ordeal continued. He was seated with the guests of honor at an elaborate breakfast, lasting for hours. Fresh dainties were brought to the table continuously, and the host never gave up trying to persuade him to taste this or that delicacy, or give his opinion of some special wine. Complimentary speeches, songs to the guitar, and choral interludes lengthened out the meal. Pale and inwardly wretched, Glückenheimer refused to be tempted.

The meal finally ended and the guests stood up for the ceremony. When it came to pronouncing the blessing on wine, the M'sader K'dushin merely touched his lips to the sacramental winecup, allowing no drop to enter them.

At last he was free to leave, amid thanks and compliments from the guests. "Now let's see if we can repay you for your great kindness," said the host. "Come with me."

They entered a room at the contents of which Glückenheimer's face brightened and his eyes gleamed. Piles of silver lay at the foot of pillars festooned with silver wreaths and garlands. Silver vases brimmed over with coin, silver trays held yet more objects of sterling silver. It was such wealth as he could not have anticipated.

"Take as much as you wish," said his companion, casually.

Glückenheimer, despite a tug at his heart, replied "Thank you for your liberality. But . . . you need not pay me. Your gratitude is enough."

"I see, such stuff as this does not appeal to you," the host responded. "Very well! We have something here that may be more to your taste." And he led the M'sader K'dushin into a chamber yet more dazzling.

The pillars were of red gold. The floor was tessellated in squares of green and yellow gold. Piles of newly minted coins, boxes of antique money, bowls and vessels exquisitely chased and figured and embossed so that the art of their craftsmanship multiplied manyfold the value of the metal, drew his fascinated gaze from one marvel to another.

"Ah, perhaps the weight concerns you?" said his host. "Have no fear of that. We shall undertake to bring it to you. You have only to choose."

Faintness welled up in Glückenheimer. He closed his eyes against the maddening sight. "No. Again I must protest that you owe me nothing," he sighed.

"This, too, is beneath you, I see. Pardon me for having misjudged you."

They entered a third room. The walls were hung with dark velvet, the floor carpeted in grey. Bowls of glowing gems stood on pedestals of alabaster. Rubies from rose pink to blood red; deep green emeralds, sapphires, diamonds like white fire, pearls of all colors and magnificent size.

But here the temptation was already less. Never had Glückenheimer imagined such splendor, and it formed no part of his dreams. It was unreal to him. He had little difficulty in saying, "No, I thank you. This is more than I have earned and I want none of it. Pray, let me return home."

THE FACE OF HIS HOST showed a grim relief. He opened another door and with his hand on Glückenheimer's shoulder led him through. This was a small, bare room with walls of whitewashed brick. Hundreds of keys were displayed hanging from plain iron hooks. Among them Glückenheimer recognized, with a last flare of astonishment, the key to his own strongbox.

His companion gave it to him, saying: "Whenever a coffer is made, two keys are fashioned for it. One is the owner's key. The other is God's. If God's key is not used, the owner will find it harder and harder to take money from the box. He will open it only to add more and more to the hoard, and, eventually, his soul will be locked up in it.

"We are spirits who serve the Lord. Remember what has been revealed to you. Take God's key, and use it."

A door on the opposite side of the little room swung open. Glückenheimer stepped through it, and found himself in his study, at home.

He burst into thanksgiving at his escape. Then he took from the box one-tenth—the Biblical tithe—of all it contained, and humbly brought it to the community chest.

●

From *Tales from the Wise Men of Israel* © 1962 by Judith Ish-Kishor. Reprinted by permission of Harper-CollinsPublishers.

The Test of Giving

Philip Zaleski

The Way of the Bodhisattva: Almsgiving.

THE CROWDS had thinned by the time I saw the beggar. He stood at the edge of the town square of Nara, that ancient city of deer parks and pilgrims just south of Kyoto. What struck me at once was his indifference to the shoppers, jugglers, and dancers who still whirled around him on this hot June day. Even when the politicians set up their sound truck, slipped on their white gloves, and began to exhort the crowd with shouts of "Banzai!" he remained still, eyes turned downwards. An enormous straw hat shielded his head, casting a long shadow over his black robe and brown sandals. In his right hand, he held a wooden begging bowl.

He was, I knew, a mendicant, a Buddhist monk who depended on alms for his livelihood. At once I decided to drop some yen in his basket. But just then a troupe of school children swarmed around me, demanding to try out their English on the bemused foreigner, and almost an hour had passed before I returned to the square. The monk stood in precisely the same place that I had last seen him—silent, still, eyes down.

I edged towards him, astonished by his capacity for remaining erect and motionless. As I approached, I noticed that his other hand was counting black prayer beads—clack, clack, clack—on a string. I pulled out some change. At least, I thought, this time I am giving to a good cause, remembering the con artist in New York City who had squeezed twenty dollars out of me with a song-and-dance about a lost wallet and a waiting wife and kids; I had felt good about that act of generosity until I saw the same man a few hours later, a few blocks away, giving another poor rube the same story. And behind that con man stood an endless chain of panhandlers, fast-talkers, drugged-out kids. How many times had I dropped some coins and walked away, or stayed to make useless conversation? Looking at the monk, I felt pleased—a bit of sorrow for his situation, a bit of smugness for my generosity, a bit of curiosity about his life. Yet something in me resisted; the money, after all, was mine. I supposed this monk was authentic, but could I really be sure, here in a strange land? And in any case, didn't he get free food and shelter from his monastery? Perhaps I should give my coins directly to the poor . . . but perhaps he was indigent. And wasn't I in a superior position when all was said and done, with piles of yen jingling in my pockets? Yet wasn't he the superior one, trusting like the lilies of the field? Wasn't his work, his apparent striving for a higher level of being, part of the equation? Wasn't that just the reason that I wanted to give him money?

I knew, without thinking it out clearly at the time, that more than a mere passage of coinage was at stake, that all this confusion and doubt had something to do with my understanding—or lack thereof—of money, of obligation, of need, of the respective place of giver and taker. Such conflicting thoughts assailed me as I dropped the yen into the beggar's bowl. He never said a word or made a sign to acknowledge my largesse (did I feel miffed by his lack of recognition? I suppose I did). So it is always with money: if not the root of all evil, at least the root of much anxiety. How poorly we rule our cash and coin; how artfully they rule us. Think of the lies, ploys, incriminations, terrors we invent in the presence or absence of money, how easily we turn our backs to truth, so that even a plain-spoken public figure like Mario Cuomo can utter, in regard to New York State's fiscal woes, such nonsense as "I don't really like money."

Might it not be possible, however, to discover a new approach to money? To understand my relation to this universal symbol of exchange in a way that brings clarity rather than confusion, perhaps even collectedness rather than tension? Above all, alms-giving is a measure of who I am, of what in me is in ascendance. Mulling over my jumbled reactions to the nameless monk, I hear again the clamor of voices that assailed me at the time, each of them expressing a part of myself—the Scrooge, the Kindheart, the Efficiency Expert, the Fare-Thee-Weller—each chirping out its patented refrain: "Go away," "Take this and go away," "Take this and prosper"—all of them spoken as if in a dream, without any real participation on my part. Is there another way of doing this, I wonder, a way of approaching someone in need so that my box of masks is cast aside, so that I truly meet the mendicant face to face on common ground? Can I learn to give a coin to another from myself *as I am*, not just as I imagine myself to be, and to him *as he is*, not just to some image I have of him? How often do I actually see the panhandler? How often do I stop, even for a moment, to

Almsgiving, then, is also a means for positioning myself in the world.

---

initiate a touch of glances or of hands? Can there be a genuine relationship at that moment? Can I learn to give, not to fulfill a social or religious code (although such codes can be forceful reminders of what I want), but through a genuine feeling of kinship, even of identity with the other?

S UCH A NEW understanding of almsgiving seems to demand a parallel understanding of the reality of levels of being, and the way in which these levels mix in every individual. To see a beggar without illusion means to see him as cloaked in rags (and, emphatically, not to deny the reality of those rags) but also to see him as cloaked in the dignity of all human beings—that is, to use the lovely Judeo-Christian formulation, as made in the image and likeness of God. In orthodox Christian terms, the giver sees Christ Himself in him whom he clothes or feeds. In Mahayana Buddhism—the teaching can be found in almost all traditions—it is said that by feeding a beggar's body with alms today, we create the conditions in which his being may be fed with the Dharma tomorrow.

As a consequence of giving alms to Christ within the beggar, the money itself changes character, turning from what Saint Paul called "filthy lucre" into a holy instrument of exchange. Instead of being in the service of my ego and its incessant demands, gold and silver come into the service of God or Dharma; that is, they lie under a different set of laws or influences, and their possibilities become unknown and wonderful. This is truly an exchange, for by sanctifying my money, I change my relationship to it and to the world that it affects. A return becomes possible, in accordance with the mysterious law by which good deeds rebound upon the doer. This is why Theravada Buddhists describe the mendicant monk as a "field of merit" (*punyakshetra*) for the giver, repayment coming in the form of an auspicious future life, and why St. Gregory the Great imagines the righteous in heaven as inhabiting houses built of golden bricks cast from the alms that they gave in life.

A LMSGIVING, THEN, is also a means for positioning myself in the world. I place my money (and thus myself, for money is a symbol, albeit not the only one, of my power) in service to what stands above or below me, depending on whether and how I save it or spend it. If I hoard my treasure, content to bask in the gleam of gold, I nourish only my ego. If, on the other hand, I circulate my money, I cultivate its living power. (Note that money circulates like blood, inflates and deflates like the lungs, and, like life itself, resists all efforts at rigid control; this is why free-market economies will always have boom and bust cycles, and control-market economies will always fail.) To participate in sacred exchange, money *must* circulate. But how? A conventional formulation might be: do I spend my money to help myself or to help others? But perhaps this is a false opposition, for by careless spending I can feed what is base in others as well as myself. Rather, I might ask whether my wealth serves the Above or the Below, God or Mammon. This suggests one way to understand Jesus' thorny teaching about the rich man and the camel: A man with mounds of gold may still pass through the needle's eye if he understands himself as steward, not owner, of the money in his grip, and consecrates it towards God's work; conversely, a poor man may get wedged in the eye if he fails to place his few coins in the service of that which will bring true benefit to himself and others. It matters how much money I have; what matters more is my relation to what I have.

Precisely because almsgiving involves tribute to what is highest in ourselves and others, we are taught that it must never be automatic, egoistical, or effortless. To the very extent that almsgiving is mechanical, it is divorced from my highest part and thus from the possibility of being directed towards another's highest part. The Gospel of Matthew therefore counsels us to leave our egos aside and to make our almsgiving a secret work:

Take heed that ye do not your alms before men, to be seen of them. . . . When thou doest alms, let not thy left hand know what thy right hand doeth: That thine alms may be in secret: and thy Father which seeth in secret himself shall reward thee openly. (Matthew 6:1, 3-4)

To act in secret, to direct my actions towards Christ within: these are difficult demands, almost impossible. They demand awareness and discernment. Ideally, these qualities will also guide the day-to-day direction and degree of giving. Only a fool will pour all his cash onto the first person who asks, for such a rash act may bankrupt (in ways both obvious and subtle) both the fool and his beneficiary. How

can I learn to gauge need? When should I give, when should another receive? As stated in the following excerpt from the Didache, one of the earliest Christian texts, proper almsgiving requires an honest appraisal by both giver and taker:

Give to everyone that asks, without looking for any repayment, for it is the Father's pleasure that we should share His gracious bounty with all men. A giver who gives freely, as the commandment directs, is blessed; no fault can be found with him. But woe to the taker; for though he cannot be blamed for taking if he was in need, yet if he was not, an account will be required of him as to why he took it, and for what purpose, and he will be taken into custody and examined about his action, and he will not get out until he has paid the last penny. The old saying is in point here: "Let your alms grow damp with sweat in your hand, until you know who it is you are giving them to."[1]

Yet how frequently I fail this test, as often by not giving as by giving too readily, letting cowardice or avarice or preconception stand in the way of clear perception. One day last summer, as my young son and I sat in a sidewalk café eating frozen yogurt, a young couple and their child entered the room. All three looked bedraggled, filthy, soaked with sweat from the terrible heat outside. They also looked scared. Instantly, the husband and father—so I took him to be— veered towards me and explained in a hoarse whisper that his car had broken down, his wife was pregnant, his child hungry, and that fifty dollars would return them to their home halfway across the state. No, I thought—a bit defensive because my son was beside me, but mostly filled with mistrust, as a parade of past encounters with rip-off artists marched across my mind. "No," I said out loud. As the young man and his family left the store, something in me leapt up—I *knew* with absolute certainty that he had been telling the truth. Yet some-

Giving to the poor in secret.

thing else—the blast-furnace heat, my little son, fear of making a fool of myself—kept me indoors. Only later that day, in conversation with a policeman, did I learn that indeed the young man had been telling the truth. My shame plagued me for days. But what I retain now isn't that shame, but the understanding—on those rare occasions when it is available to me—that each like situation must be measured as carefully and sincerely as possible.

BEING ASKED for money, then, is always a test: one in which the worthy response is not always to give, but always to focus whatever attention I possess on him who asks, and on myself, and on the money that joins us. The supreme form of such testing may well be the tithe, a traditional practice that involves assigning ten per cent of income to the religious community. The practice is old, being mentioned several times in the Pentateuch ("Of all that thou shalt give me, I will surely give the tenth unto thee"—Genesis 28:22). It is also controversial, usually because, in ancient Israel and elsewhere, tithing was obligatory. No doubt this runs counter to popular ideas about freedom and fair play—the degree of resistance that the mere mention of tithing evokes nowadays is, in fact, quite extraordinary, and says much about our relation to our money—but to a large extent the issue of obligation is a red herring. Tithing was imposed by law during an era of theocracies, and in that context it served more or less the same function as our annual income tax. There is one distinction, however—the Bible describes the ancient Israelites as making their tithe with great joy:

And proclamation was made throughout Judah and Jerusalem, to bring in for the Lord the tax that Moses the servant of God laid upon Israel in the wilderness.

And all the princes and all the people rejoiced and brought their tax and dropped it into the chest until they had finished. (2 Chronicles 24:9-10)

Why the joy? Because if alms is money consecrated to that which is highest within ourselves, then the tithe is money consecrated to the outward manifestation of that highest reality. In the instance above, the funds are directed towards the restoration of the Temple; it should be noted, however, that tithing is still practiced today in some churches, on the same voluntary basis as its more familiar analogues—the weekly collection plate, annual dues, and so on—whatever means are necessary to support the material underpinnings of the spiritual search. Here lies a deep truth about the sacred circulation of money, be it almsgiving or tithing: the issue is not obligation but responsibility, not enforcement but valuation. Faced with the prospect of restoring the house of God, the Israelites felt no hesitation. (This explains, by the way, why almost all traditions, even those most committed to helping the destitute, adorn their sacred spaces with fabulous wealth; it is a matter of valuation and a recognition of levels, bejeweling the abode of God being a creative manifestation of the same impulse that, in its submissive form, demands that we remove our shoes on holy ground.) Tithing and its corollaries can perhaps be best understood as a weighing of values, an attempt to find one's place in the cosmic network; it is also a measure against which one strives, against which one balances the other demands of life—paying the rent, financing the car, going on vacation. A parent with hungry children may not pay his or her tithe, but the failure to do so may bite, and that bite itself may serve as a reminder of why the tithe is asked, the reason for the existence of the sacred organization—be it Christian church, Islamic *tariqa*, Buddhist *sangha*, Jewish synagogue, Hindu ashram.

TITHING MAY EVEN BE an entry into the idea of payment: that nothing is free in the economy of the cosmos, that we must give for what we get. This is a great mystery, not easily discussed, but perhaps it can be said that in the lives of great spiritual figures we see lessons in universal economy incarnated—for instance, in the example of Christ, who according to orthodox thought became himself Holy Coinage, paying for the Cosmos with his own flesh and blood, a payment reenacted each week in the Eucharistic feast.

Much remains unexamined (why, for instance, do monks in so many traditions take a vow of poverty?) and almost all remains unexplained. Yet when I look back on that moment with the monk in the courtyard in Nara, I realize that in that moment—just as, in any moment of my life—I find it all before me: my sniveling, selfish attitudes towards money, others, and myself; the call of a higher voice; the instant counter-call of the lower; the need to situate myself in the field, to understand why I am here and what I should do. Money, it seems, can be a teacher as well. ●

NOTES

1. Quoted in *Early Christian Writings*, translated by Maxwell Staniforth (Harmondsworth: Penguin Books, 1968), p. 191.

Mother Teresa and the Poorest of the Poor

Poverty and homelessness, the dark side of the theme of this issue, are the battleground of some of today's most heroic figures —Mother Teresa and the Missionaries of Charity, the order she founded. In Mother Teresa, *an 82-minute feature documentary film by Ann and Jeanette Petrie, we hear Mother Teresa speak about poverty and service. In one scene when the narrative moves from Calcutta to New York, she is heard to say:*

"You can find Calcutta all over the world if you have eyes to see. Every place, wherever you go, you'll find people unwanted, unloved, uncared, rejected by the society, completely forgotten, completely left alone. That is the greatest poverty of the rich countries."

■ ■ ■

A little further on, at a Harvard-Radcliffe graduation ceremony where she herself has just received an honorary degree, Mother Teresa speaks again on this same theme.

"You will of course ask me, 'Where is that hunger in our country? 'Where is that nakedness in our country? Where is that homelessness in our country?' Yes, there is hunger—maybe not the hunger for a piece of bread, but there is a terrible hunger for love. We all experience that in our lives: the pain, the loneliness. We must have the courage to recognize the poor we may have right in our own family. Find them, love them, put your love for them in living action, for in loving them you are loving God himself. God bless you."

■ ■ ■

As witnessed throughout the film, the sisters take vows of obedience, chastity, poverty, and a fourth vow of whole-hearted free service to the poorest of the poor. Mother Teresa is very strict about the observation of the vow of poverty in their service to the poor. After a scene where the sisters have stripped down their new residence to the barest essentials, we hear:

"Many people don't understand why we don't have a washing machine, why we don't have fires, why we don't go to the cinema, why we don't go to parties. These are natural things and there's nothing wrong in having them, but for us, we have chosen not to have. For us to be able to understand the poor, we must know what is poverty. We accept no government grants, no church maintenance, no salary, no fees, no income as such. But the flowers and the birds, they don't do anything, yet God takes care of them. And he takes care of us, we are more important than the birds.

"And in spite of our weakness and our poverty, there is no worry for us because we know somehow, some way, things will be all right."

■ ■ ■

It is clear that she does not allow any kind of fund-raising. In response to a man who wishes to organize a campaign to raise money for her, she says:

"I don't allow anybody to raise money nor ask in our name or any-thing. You yourself, you can help, but please don't ask in our name. I'm not allowed to ask any fund-raising.

"We want to depend solely on Divine Providence. That people give us until it hurts them, to share in the joy of loving."

■ ■ ■

Later in the film, during a press conference, she is asked how she feels about being called a "living saint." Her response:

"You have to be holy in your position as you are, and I have to be holy in the position that God has put me. So it is nothing extraordinary to be holy. Holiness is not the luxury of the few. Holiness is a simple duty for you and for me. We have been created for that." ●

Excerpts from Mother Teresa, *a film by Ann and Jeanette Petrie, © 1987, with permission of Petrie Productions. Also available on videocassette.*

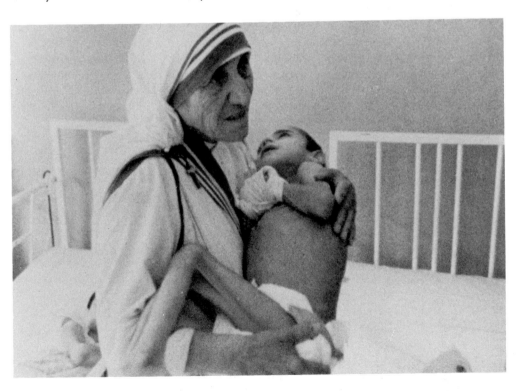

From Rhinegold to Ring and Back Again

Rob Baker

RICHARD WAGNER'S four-evening operatic cycle, *The Ring of the Nibelung*, is a monumental, highly complex epic based on a theme of profound simplicity: the corruption of divine energy (symbolized by a glowing source of golden light under the waters of the Rhine River) when greed and lust debase the Rhinegold into something else entirely: a Ring of Power for the acquisition of profane wealth and the domination of mankind. What begins as a neutral force resting softly in pure, primal waters in the prologue of *Das Rheingold* will end up as the generator of complete, cosmic destruction: *Götterdämmerung (The Twilight of the Gods)*. Yet even though the gods themselves, who fall victim to the compulsive curse of the Ring, must be destroyed, the divine impulse —far purer than they can ever be— lives on, to light the way to a new order, freer and higher than the old one, which barely managed to keep its spark alive.

Wagner's *Ring*, in other words, is all about money—about how the universe extracts payment, even from the gods who try to rule it; about values,

real and imagined; about mundane coinage and divine wealth; about the Gold that glues the cosmos together and the Ring that entraps it instead, grasping it greedily until its innate energy can no longer be contained and explodes in complete destruction.

In our own lives, though, what is money, really? What does it symbolize, represent—not just in the outside world, but within? Where do we find the Rhinegold there, instead of being trapped by the Ring in all its various, contemporary manifestations?

And why is it that money touches such a chord in our being, one that is, most often, so dissonant, so distressing to the ear and the heart, so difficult to harmonize?

We blame money—or our lack of it—for almost everything, all the while secretly insisting that if we only had it, everything would be all right, all our problems would be solved. But money ends up being neither savior nor villain; its absence or presence simply puts us—our very being—into question again and again, and that questioning most often entails suffering, self-examination, and sacrifice.

DESPITE HIS OWN legendary problems with money—and his exceptionally crass attitudes about payment and responsibility—Richard Wagner nevertheless managed to make *The Ring* a rich and moving examination of the layers and levels of meaning behind this theme of money: where its power comes from, what happens when that power is ignored or abused, the need for its proper re-alignment or restoration to its source. Wagner's genius was that, as the author of both the words and the music for *The Ring*, he recognized the subtle, interconnecting power of various traditional symbols. It is necessary to look at his musical and poetic images in some detail to see this: not only the basic touchstones, the Rhinegold and the Ring, but also the related emblems and ideas of Light and Darkness, Sleep and Awakening, Love's Renunciation and Love's Redemption, the Spear of Law and the Sword of Need—for a fuller understanding of the work's many complexities.

All these themes are stated powerfully in the first opera of the cycle, *Das Rheingold*, which Wagner described as a prologue to the other three.

The first scene of *The Ring* thus begins in darkness, as the orchestra plays a single chord in E-flat major. The chord continues unbroken for 136 bars (over four minutes) before anything starts to happen onstage. Musically, the chord grows from its original three notes to include various other harmonics, then little arpeggios, which rise and fall like waves—for this prologue to *Das Rheingold* does indeed take place under the waters of the Rhine River. Most commentators have stressed the calming, peaceful effect of this long, sustained chord, but a careful ear senses something a bit sinister and foreboding about it as well: there's an edge to it, a touch of darkness. Na-ture—in this scene and in life—may not be quite the blissful, orderly, and harmonious place we romantically envision; it is, rather, a neutral state in which all things are potential, but as yet unrealized.

Next comes Light—dimly at first, then growing progressively brighter, in waves, like the sound. We eventually make out three mermaid-like creatures, the Rhinemaidens, who first sing a kind of melodic primal gibberish ("Weia! Waga! . . . Wagala-weia! Wallala weiala weia!").

A dark, sluggish creature appears: Alberich the Dwarf, an earth-bound, ugly, clumsy toad-like being, slipping in the slime, unable to swim. He is attracted to the grace of the Rhinemaidens, the very lightness of their being. He wants to play, but his idea of play is heavy-handed, grotesque, inspired only by physicality and lust. They tease and torment him with their easy agility; Alberich grows more and more frustrated and angry.

At last, the maidens and the score draw attention to the secret which the trio is guarding: the Rhine-gold, an essential, natural source of power and light, nestled in the depths of the Rhine and reflecting the higher light of the Sun:

Look sisters!
In the depths it awakens with smiles. . . .
Look, it smiles
in the bright light.
There through the waters
flows its shining ray.
Heiayaheia!
Heiayaheia!
Wallalallalalaa leeiayahei![1]

It's critical that the Rhinegold is not seen as a hoard of natural treasure, but more as an uncorrupted energy, a symbol of true value or worth, a force "that sleeps and then wakes" (and can awaken human sleepers as well)—a magical source of marvelous powers:

"A man could win the world's wealth if he forged a ring from the Rhinegold. He could gain limitless power."[2] But there's a catch: "Only a man who would renounce love, who can forswear the ecstasies of love forever" can gain the Gold.

Alberich has little trouble deciding: "Make love in the darkness, you brood of fishes. I'll cut off your light. . . . Now do I curse love!"

This announcement is accompanied by a musical leitmotif or melodic signature, a device that Wagner uses throughout all his later operas to call up the emotional resonance of a symbol, a character, or an idea. The "Renunciation of Love" motif, in its melancholy, minor-key melodic power, is one of the most affecting in the entire *Ring*, representing, as it does, the very crux of the work's meaning: by renouncing true feeling (not romantic sentimentalism, not physical lust, but *real* feeling), man steals the source of divine energy and power, corrupting it, debasing it, physicalizing it into heaviness, darkness, materialism, and greed: the Ring of power, domination, and death, the Ring that will corrupt and destroy not only Alberich, but everyone who comes in touch with it, even Wotan, the chief god of the pantheon of Scandinavian mythology.

The music conveys the same message. Wagner makes the leitmotifs, or leading motives, as they are perhaps more properly called, grow out of each other and comment on each other; as critic Deryck Cooke, among others, has pointed out in his study, *I Saw the World End*,[3] musical transformations almost always underpin emotional or thematic transformations in the work. Almost all the major themes grow out of the chord and wavelike arpeggios representing Nature in the Prologue, including both the joyful melodic hymn to the Rhinegold

(which is later transformed into a woeful dirge, the lilting "*Heiayaheia!*" turning to the guttural "*Weh! Weh!*"), and the powerfully contrasting theme of the Ring that the Gold is to become. Instead of the rhythmic liveliness and melodic simplicity of the Gold leitmotif, the Ring motif is slower and darker, in a minor instead of a major key. It is full of sharps and (as Cooke describes it) "complex chromatic dissonance," built on a "sinister harmonic basis" which will undergo further transformation throughout the cycle to denote the scheming, resentment, and "progressive disillusionment" associated with attachment to the Ring.[4]

But almost immediately after its first statement, in the scene with Alberich, the Ring motif undergoes a rather striking transformation, becoming a bridge to the next scene, a vista of Valhalla, the just-constructed new home of the gods. By switching back from minor to major key, Wagner makes the "sinister harmonic basis much more genial,"[5] so that the cadence takes on a kind of majestic pomposity, denoting, in this case, a lust for power on a somewhat higher level than Alberich's materialistic greed. The stage darkens, then is bathed in full sunlight, revealing Wotan—asleep.

WAGNER'S GODS—indeed, the gods of Scandinavian mythology, in general—are not to be taken as divinities: though slightly higher up the ladder than mere men, they are distinctly anthropomorphic—with most of the shortfalls and shortcomings of humans. Wotan, though he heads the pantheon, is hardly a paragon of virtue or humility: he's unfaithful, constantly, to his wife Fricka, and he's as vain and egotistical and greedy and insensitive as anyone in the

operas except, perhaps, Alberich and his son and accomplice, Hagan. And he's also, quite tellingly, asleep when we first see him.

"Wotan, wake up," pleads Fricka, as the scene shifts to the home of the gods.

"Man's honor and might will be renowned forever," he dreams on.

"Awake and think!" insists Fricka, urging him to leave his deceptive dreams behind.

He opens his eyes, only to look upon the supreme manifestation of his ego: Valhalla, which he's commissioned two giants, Fasolt and Fafner, to build for him.

Fricka reminds him, "Have you forgotten the price that must be paid . . . the tribute that must be rendered?" Wotan had promised her sister Freia, guardian of the Golden Apples which keep the gods young, to the giants in exchange for their labors.

Since Freia is the goddess of love, Wotan, too, has forsworn love in a very real way by his rash bargain. But the pompous god merely shrugs. "As for the contract, don't worry about it." But the leitmotif of his Spear—a branch he wrested from the World Ash Tree at the center of the universe (wounding it fatally, as we—and he—will learn later) which is now the symbol of the Law, of oaths and contracts in general—sounds as a monumental descending scale: the higher brought lower, involution with its constant danger of corruption by the ordinary.

Only the appearance of Loge, the trickster who is half-god and half-human and represents flickering, tricky, elemental Fire, saves the day. He knows of Alberich's theft of the Ring and suggests that Wotan steal it from Alberich (who, after all, stole it himself, Loge reasons) and offer it to the giants instead of Freia. The giants—and Wotan—agree.

Loge's characterization as sly and deceitful is linked to his association with Fire in its more negative aspects. Though it can cleanse and purify and regenerate, Fire also can destroy, rage, run out of control. Unlike the pure, steady light of the Sun (the higher, the divine, the center), or the corresponding reflective light of the Rhinegold (divine energy transmitted on the human plane, with the promise of the power to do or make something in a significant way), Loge is linked to the flashing, flaming, unpredictable, dangerous aspect of Fire—a different sort of light entirely. It is Loge's flames that have allowed Alberich to forge the Rhinegold into the Ring: a symbol not of glimmering possibility and lightness of doing, but of the crass cycle of heavy-handed manufacturing, materializing, and hoarding rather than using.

L OGE LEADS WOTAN to Alberich, and the two trick him out of both his hoard and the Ring. Wotan, already realizing that the "magic hidden inside the Gold" may still be present in the Ring, has designs on it for himself: he has no intention of giving it to the giants, even to win back Freia (whom they are holding hostage). Fricka, too, has her own agenda: Could this Ring insure a husband's fidelity, she wonders. Wotan's theft of the Ring also summons forth a curse: the darkest leitmotif of the operas, a heavy but rhythmic descending bass melody reminiscent of that of Wotan's own broken-treaty Spear. "You delude yourself, dreaming on your marvelous heights," Alberich snarls, when he first realizes what Wotan has in mind. "You immortals despise us black gnomes. Beware! Bewitched by gold, the lust for gold will enslave you! Beware the armies of the night when we Nibelungs swarm into daylight! . . . Take heed, arrogant god. If I sinned, I sinned only against myself. But you, immortal one, sin against all that ever was or will be if you ruthlessly steal my ring!" Alberich then delineates the curse:

Its gold granted unlimited power.
Now let its magic bring death to the
    owner.
No man shall gain joy from it.
Anguish will consume whoever pos-
    sesses it.
Everyone will lust after its power,
But no one shall have pleasure from it.
Its owner shall reap no gain.
It will bring him only death.
The lord of the Ring
Shall be the Ring's slave.

Wotan offers the giants Alberich's hoard, but they balk, insisting that he add the Ring to the pile. Wotan refuses and Erda, the Earth goddess, appears, telling him to "yield" (though the German, "Weiche, Wotan, weiche!" no doubt intentionally echoes Fricka's opening "Wotan! Erwache!", "Wake up, Wotan!"). Finally, the god does—very grudgingly—come to his senses and gives the Ring to Fafner. Fasolt then grabs it from Fafner, and Fafner immediately kills his own brother, without the slightest hesitation—to the tune of the first replay of the leitmotif of the curse.

Freia and her apples are saved. The giants get their payment—and more. Wotan is already dreaming again, this time of a human hero, a free man, who will, he thinks, be able to rescue the Ring from its curse with a magic Sword and save the gods from destruction. The gods themselves are similarly unaware of what is really unfolding: they are ready to go across a rainbow bridge to Valhalla, their new home. Yet all is hardly sweetness and light. "A heavy mist hangs in the air," says Donner, the thunder god. "I'll gather the clouds together in a thunderstorm and clear the air." He does his best. Then Wotan pompously announces:

The last light of the sun gleams.
In this glorious glow, the fortress shimmers brightly.
This morning it glittered bravely, without a master.
Now it proudly beckons to me.
Between morning and evening
It was won through suffering.

Loge can barely restrain his laughter at Wotan's wishful thinking, realizing that the vainglorious god has barely even begun to taste the suffering that will be his lot in the next thirteen hours. Though he touched the Ring only for a few minutes, its curse will haunt him until *das Ende*, the end of everything, for which he will soon be longing.

In the distance, the Rhinemaidens are heard, begging for the return of the Gold. Wotan complains about their wailing. Loge intervenes: "You in the waters, stop your crying. Hear Wotan's wish. Stop seeking the Gold, you maidens. Bask instead in the newfound glory of the gods."

Wotan's wish will get him into a good deal more trouble before the cycle is over, but for now, the gods stride happily into Valhalla, heedless of the Rhinemaidens' final warning about the Gold: "Only in the deep is it trusted and true. All that rejoices above is cowardly and false." When the Ring is held to be higher than the Gold, the pretended hierarchy is not only false, but doomed.

IN THE FOLLOWING three operas — *Die Walküre, Siegfried* and *Götterdämmerung* — the curse continues to stalk and haunt everyone who touches the Ring or even crosses the path of those greedily pursuing its power.

Central to all the convoluted developments of the plot are two major conflicts: between Wotan's rigid Will and the very different willfulness of his daughter, the virgin Valkyrie-warrior Brünnhilde; and between Wotan's Spear and another weapon of his creation, the Sword.

It is Brünnhilde's challenge of Wotan that is central to the second opera, *Die Walküre*: Insisting that she understands the true will of her father's heart better than he himself does, she defies him in battle and, for her trouble, gets banished to a ring of fire (note the ring, note the fire) on a mountaintop. The authority of Wotan's Spear (the descending-scale motif of involuting Law) must be challenged by the magic Sword (heralded by an octave-leap down, instead of the note-by-note descending scale of the Spear, and then by a brassy ascending melody). The Sword, too, was originally created by Wotan but, importantly, it is now out of his hands and control so that a free man, a human not allied directly to the gods, can wield it.

The motif of the Sword again grows directly out of the motifs of Nature and the Gold. It is, in a sense, the Gold rightfully used, not hammered into a Ring, a closed circle of corrupted Power, endlessly repeating itself, but rather into a pointed weapon to initiate action, to do. The Sword is named *Notung* in German: Needfulness, or essential necessity, a deep and powerful sense of lack. It generates what must be done to return the Gold to its source, which includes shattering the Spear of the old law which was corrupted by Wotan's ego.

The tragic hero who will bear the Sword is Siegfried, whose innocent nature leaves him free to challenge both symbols of corrupted power, human and divine: the Ring and the Spear. He thus becomes the agent of transformation who not only wrests the Ring from the dragon, but also awakens Brünnhilde, still sleeping in her ring of fire. Only through

Brünnhilde—the other essential ingredient in this alchemical marriage and arguably the real protagonist of the whole cycle—can he really redeem the original Gold through the very Love that was forsworn to allow the forging of the Ring in the first place.

To Siegfried's gentle "*Erwache*," Brünnhilde awakens, greeting the Sun, the Light, the "radiant day," before turning to the hero who has caused her to stir from her sleep at last.

For a moment the Ring is aligned with the Sun, with Light, with Awakening, with the Gold of its original state. Love, which Alberich renounced, is possible again, through Siegfried's innocence and freedom and Brünnhilde's devotion to the real essence of divine will—Wotan's higher wish, which even Wotan, obviously, couldn't articulate. "I am wise because I love you," she tells Siegfried, then adds: "I alone understood Wotan's purpose. I battled and struggled for it. I even defied the one who conceived of it. I did not think. I only felt. And what was that purpose? To me, it meant loving you."

But Siegfried, even in his innocence, has touched the Ring. Instead of the higher Love Brünnhilde has in mind, he wants earthy, carnal fulfillment—one more in tune with yet another meaning of Fire (passion, lust, carnal desire) than with the Sun (the higher, divine aspect of Love): "Awake!" he says, speaking to a different part of Brünnhilde's nature than before, "Be a woman to me."

"The man who awoke me now wounds me," she sighs. "I am no longer Brünnhilde." But she weakens: "My senses are reeling. My wisdom fails me." The Ring is now close to her own finger. She speaks in images of night, fog, gloom: "The sun glares on the day of my disgrace."

Siegfried, the ringbearer, gets his way; Brünnhilde gives in to his advances. The pull of the Ring, the token forged from Earth and Fire, of materiality and sensuality, grows stronger, blotting out all memory of higher Air and purer Water, as Brünnhilde submits: "My godly serenity is lost in surging desire. . . . Divine wisdom flies, chased by love's conquering cry."

The Ring's final curse unfolds with a deadly insistence in *Götterdämmerung*, the most dramatic of the cycle's four operas. Siegfried and Brünnhilde are separated; Siegfried, drugged by a potion, renounces Brünnhilde and marries another woman; Brünnhilde (who has touched the Ring herself by now) seeks revenge, aided by Alberich's son Hagan; Siegfried is killed.

At the end, there is a majestic funeral procession and Brünnhilde enters, takes the Ring again from Siegfried's finger and prepares to meet her own death. "This truest of men had to betray his wife so that she could gain wisdom." She lights Siegfried's funeral pyre on the banks of the Rhine and prepares to walk into the flames, wearing the Ring. This time the fire will purify the curse at last, returning the Ring to its source when the river overflows its banks and reclaims the Gold. Above in Valhalla, the failed gods simultaneously meet their end, also in a blaze of light. "The blazing fire grips my very heart," Brünnhilde says, then walks into the flames, the Ring is transformed back into Gold, the gods' ego-monument crumbles, and, slowly, a new Sun, a new order, a new cosmos begins to grow, one in which, perhaps, Love will not be forsworn, real values can flourish, and a balance of the elements can obtain.

THE REACTIONS of characters in *The Ring* to the power of money is thus intricately tied to how they individually and collectively relate to the dual symbol of the Gold and the Ring. Most of the reactions are emotional and predictable: greed, lust, anger, even ignorance and denial, in the case of Siegfried, who learns all too soon that that attitude towards money has no more validity than cupidity and overidentification.

Perhaps the relationship to money is rightfully (needfully, as the name of Siegfried's sword indicates) painful and difficult: a struggle that can only be resolved by facing up to it boldly, with all our wits about us. In the entire cycle, only Brünnhilde recognizes this and can redeem the Gold, which she does at the very end of *Götterdämmerung* in a scene of considerable dignity and power. Standing in front of Siegfried's funeral pyre, she sings:

Now I take up my inheritance.
Accursed ring, terrible ring,
I take your gold and now I give it away.
Wise sisters of the water's depths,
you swimming daughters of the Rhine,
I thank you for your good counsel.
I give you what you crave:
from my ashes take it for your own!
The fire that consumes me
shall cleanse the ring from the curse!
You in the water, wash it away
and keep pure the gleaming gold
that was disastrously stolen from you.[6]

Like Wotan, Brünnhilde touched the Ring only briefly, and even though her wish, her wisdom, her impulse in contacting it were higher than his in both thought and feeling, the Ring demanded payment, and that demand could not be fully met with denial, anger, or even love on the physical level; it could only be accepted, and that acceptance allowed transformation and transcendence: the redemptive power of higher Love—enlightened, needful, and awake. ●

NOTES

1. English translation by William Mann © 1989 by William Mann. From the libretto to the Bernard Haitink recording of *Das Rheingold*, EMI Records, Ltd., Hayes, Middlesex, England.
2. All remaining quotes cited from the libretto, other than the two otherwise noted, are from the subtitles of the *Ring of the Nibelung* production by the Metropolitan Opera of New York City, as broadcast on the Public Broadcasting System in June, 1990.
3. Cooke used much of the same material for his classic sampler of the *Ring* leitmotifs, which was originally packaged as part of the George Solti recordings of the cycle for Decca/London Records. Cooke's audio commentary is no longer available as part of the Solti set, but it is available on audio cassette as *Weaving the Ring* from the Wagner Society of New York, P.O. Box 949, Ansonia Station, New York, NY 10023, © 1989 by the Wagner Society of New York.
4. *Ibid.*
5. *Ibid.*
6. English translation by Lionel Salter, © Lionel Salter, from the Karl Böhm recording of *Götterdämmerung*, made at the Bayreuth Festival in 1967 and issued that year by Philips Records.

Payment

P. D. Ouspensky

From Ouspensky's account of working with G. I. Gurdjieff in St. Petersburg before the Russian Revolution.

MANY PEOPLE were very indignant at the demand for payment, for money. In this connection it was very characteristic that those who were indignant were not those who could pay only with difficulty, but people of means for whom the sum demanded was a mere trifle.

Those who could not pay or who could pay very little always understood that they could not count upon getting something for nothing, and that G.'s work, his journeys to Petersburg, and the time that he and others gave to the work cost money. Only those who had money did not understand and did not want to understand this.

"Does this mean that we must pay to enter the Kingdom of Heaven?" they said. "People do not pay nor is money asked for such things. Christ said to his disciples: 'Take neither purse nor scrip,' and you want a thousand roubles. A very good business could be made of it. Suppose that you had a hundred members. This would already make a hundred thousand, and if there were two hundred, three hundred? Three hundred thousand a year is very good money."

G. always smiled when I told him about talks like this.

"Take neither purse nor scrip! And need not a railway ticket be taken either? The hotel paid? You see how much falsehood and hypocrisy there is here. No, even if we needed no money at all it would still be necessary to keep this payment. It rids us at once of many useless people. Nothing shows up people so much as their attitude towards money. They are ready to waste as much as you like on their own personal fantasies but they have no valuation whatever of another person's labor. I must work for them and give them gratis everything that they vouchsafe to take from me. 'How is it possible to *trade in knowledge? This* ought to be free.' It is precisely for this reason that the demand for this payment is necessary. Some people will never pass this barrier. And if they do not pass this one, it means that they will never pass another. Besides, there are other considerations. Afterwards you will see."

The other considerations were very simple ones. Many people indeed could not pay. And although in principle G. put the question very strictly, in practice he never refused anybody on the grounds that they had no money. And it was found out later that he even supported many of his pupils. The people who paid a thousand roubles paid not only for themselves but for others. •

Excerpted from In Search of the Miraculous: Fragments of an Unknown Teaching *by P. D. Ouspensky, © 1949 and renewed 1977 by Tatiana Nagro. Reprinted by permission of Harcourt Brace Jovanovich, Inc.*

EPICYCLE

ROBIN A. SMITH

The Simple Grasscutter / Indian

FAR AWAY AND LONG AGO, an old man lived alone at the edge of a jungle, earning his living by cutting grass for fodder. He lived a simple life and saved most of his meager income in a pot. One night, he wondered how much money he had accumulated through the years, and when he looked in his pot, he was astonished to discover a goodly sum. He rubbed his head, wondering what to do with his small fortune. He did not desire more than what he already possessed, nor did he want the pennies to sit unused. Then he had an idea.

The next day the old man went to a jeweler, and brought a beautiful gold bracelet. He visited a friend who was a merchant and asked, "Who is the most beautiful and virtuous woman in the world?"

The merchant immediately answered, "It is the Princess of the East!"

"Then take this bracelet to her," the old grasscutter said, "and tell her only that it is a gift from one who admires beauty and virtue."

The merchant left on one of his many voyages and soon enough came to the palace of the Princess. He gave her the bracelet and the old man's message. The Princess was delighted by the lovely gift, but puzzled by the anonymity of the giver. She sent a present in return anyway—bolts of precious cloth. The merchant returned to the grasscutter and presented him with the rolls of silk.

The old man did not know what to do with the precious cloth and had no desire to keep them. So he asked his friend, "Who is the most handsome and virtuous man in the world? Surely he can use these more than I!"

"That would be the Prince of the West," the merchant replied, and at the old man's request, he took the precious cloth to the noble young man. The Prince was perplexed by the anonymous gift, but he gave the merchant a dozen thoroughbred horses in return.

"What am I to do with them!" the old man exclaimed when his friend presented him with the horses. Then the grasscutter had an idea. "Keep two horses for yourself, my friend, and take the rest to the Princess of the East. She will care for them better than I can."

The merchant laughed, and travelled to the east, giving the horses to the Princess. Perplexed by the new and even more generous gift, she asked her father what to do. "Return a present so magnificent," he said, "your admirer will be ashamed to send anything more! And that will be the end of the matter." So the Princess entrusted the merchant with twenty mules, each loaded with silver.

"Twenty mules all with silver!" the old man exclaimed when his friend returned. "What am I to do with them?" So the old man asked his friend to keep six of the animals with all their treasure, and take the rest to the Prince of the West.

When the Prince saw the sumptuous gift, he was astonished. Not to be outdone, he sent back twenty camels and twenty elephants, each loaded with precious gifts. When his friend returned with the caravan the grasscutter exclaimed, "There is only one thing to do!" And the old man asked the merchant to keep two camels and two elephants, and then lead the rest to the Princess of the East.

When the merchant arrived at the palace of the Princess, her father and mother were astonished at the magnificent boon. "This can only mean one thing," the King and Queen told the Princess, "the man wishes to marry you!" So they asked the merchant to conduct them to their daughter's mysterious suitor. The merchant was horrified. If the King and Queen discovered a poor old man was the cause of the whole affair, they would be furious! But the merchant could not refuse the King's request, and he glumly led the royal party on a journey back home.

A day away from the old man's village, the merchant rode ahead to warn his friend. "What are we to do?" the merchant wailed. Then together he and the grasscutter bemoaned their fate, fearing the wrath of the King. Later that night, the merchant returned gloomily to the royal party, and the old grasscutter went to a cliff. It was dark by then, and he wanted to throw himself over the edge. "Better to die by night, than be humiliated in the morning!" he thought. But try as he might, he was too terrified to jump. In utter shame, he collapsed on the ground.

A LIGHT then shone from below the cliff, and two angelic beings appeared. "Why do you despair?" they asked the old man, and he explained his situation. The luminous beings smiled, and one of them touched the old man, while the other waved at his hut. In an instant, the old man's rags were changed into rich robes and in place of his shanty there sprang up a magnificent palace. Then the beings vanished.

The old man was dumbfounded and wandered about the palace in confusion. Servants immediately rushed to him and brought him inside, putting him to bed in soft cushions. The next morning, they woke him and dressed him, just as trumpeters sounded a greeting from the palace ramparts. The King and Queen of the East had arrived with their daughter! And leading them was a very perplexed merchant, staring in bewilderment at the new palace. The grasscutter greeted the royal party, while his servants set out a magnificent feast. Then the King approached the old man.

"Do I understand that you wish to marry my daughter?" the King asked. "If so, I heartily agree!"

The old man was astonished, but he declined. "I am too old for her," he said. Then he had an idea. "But I know who will be the perfect husband!" he exclaimed, and so he asked the merchant to invite the Prince of the West for a visit. The Prince arrived and fell in love with Princess. So the two married in the old man's palace, and everyone celebrated for days on end. Finally the King and Queen departed, and the Prince and Princess, too. But as they rode off, they paused to salute the old man. The King's soldiers sounded a hundred trumpets, and waved a thousand flags. The old man smiled, standing outside his new home. And he bid them all a fond farewell, waving a frond of the freshest and sweetest grass he had gathered that very morning. ●

Towards A Philosophy of Wealth

IT WOULD BE WELL also to collect the scattered stories of the ways in which individuals have succeeded in amassing a fortune; for all this is useful to persons who value the art of getting wealth. There is the anecdote of Thales the Milesian and his financial device, which involves a principle of universal application, but is attributed to him on account of his reputation for wisdom. He was reproached for his poverty, which was supposed to show that philosophy was of no use. According to the story, he knew by his skill in the stars while it was yet winter that there would be a great harvest of olives in the coming year; so, having a little money, he gave deposits for the use of all the olive-presses in Chios and Miletus, which he hired at a low price because no one bid against him. When the harvest-time came, and many were wanted all at once and of a sudden, he let them out at any rate which he pleased, and made a quantity of money. Thus he showed the world that philosophers can easily be rich if they like, but that their ambition is of another sort.[1]

—Aristotle

IT IS NOT the man who has too little who is poor, but the one who hankers after more. What difference does it make how much there is laid away in a man's safe or in his barns, how many head of stock he grazes or how much capital he puts out at interest, if he is always after what is another's and only counts what he has yet to get, never what he has already. You ask what is the proper limit to a person's wealth? First, having what is essential, and second, having what is enough.[2]

—Seneca

DESIRES ARE only the lack of something: and those who have the greatest desires are in a worse condition than those who have none or very slight ones.[3]

—Plato

WHAT IS LONG established seems akin to what exists by nature; and therefore we feel more indignation at those possessing a given good if they have as a matter of fact only just got it and the prosperity it brings with it. The newly rich give more offense than those whose wealth is of long standing and inherited. The same is true of those who have office or power, plenty of friends, a fine family, etc. We feel the same when these advantages of theirs secure them others. For here again, the newly rich give us more offense by obtaining office through their riches than do those whose wealth is of long standing; and so in all

other cases. The reason is that what the latter have is felt to be really their own, but what the others have is not; what appears to have been always what it is is regarded as real, and so the possessions of the newly rich do not seem to be really their own.[4]

—Aristotle

SOCRATES. There seem to be two causes of the deterioration of the arts.
ADEIMANTUS. What are they?
Wealth, I said, and poverty.
How do they act?
The process is as follows: When a potter becomes rich, will he, think you, any longer take the same pains with his art?
Certainly not.
He will grow more and more indolent and careless?
Very true.
And the result will be that he becomes a worse potter?
Yes; he greatly deteriorates.
But, on the other hand, if he has no money, and cannot provide himself with tools or instruments, he will not work equally well himself, nor will he teach his sons or apprentices to work equally well.
Certainly not.
Then, under the influence either of poverty or of wealth, workmen and their work are equally liable to degenerate?
This is evident.
Here, then, is a discovery of new evils, I said, against which the guardians will have to watch, or they will creep into the city unobserved.
What evils?
Wealth, I said, and poverty; the one is the parent of luxury and indolence, and the other of meanness and viciousness, and both of discontent.[5]

—Plato

P ROPERTY SHOULD BE PRIVATE, but the use of it common; and the special business of the legislator is to create in men this benevolent disposition.
. . . How immeasurably greater is the pleasure, when a man feels a thing to be his own; for surely the love of self is a feeling implanted by nature and not given in vain, although selfishness is rightly censured; this, however, is not the mere love of self, but the love of self in excess, like the miser's money; for all, or almost all, men love money and other such objects in a measure. And further, there is the greatest pleasure in doing a kindness or service to friends or guests or companions, which can only be rendered when a man has private property. . . . No one, when men have all things in common, will any longer set an example of liberality or do any liberal action; for liberality consists in the use which is made of property.[6]

— Aristotle

M ONEY HAS never appeared to me as valuable as it is generally considered. More than that, it has never even appeared to me particularly convenient. It is good for nothing in itself; it has to be changed before it can be enjoyed; one is obliged to buy, to bargain, to be often cheated, to pay dearly, to be badly served. I should like something which is good in quality; with my money I am sure to get it bad.[7]

— Jean-Jacques Rousseau

A DOLLAR is not value, but representative of value, and, at last, of moral values.[8]

— Ralph Waldo Emerson

T HE MONEY which a man possesses is the instrument of freedom; that which we eagerly pursue is the instrument of slavery.[9]

— Jean-Jacques Rousseau

M ONEY IS like muck, not good except it be spread.[10]

— Francis Bacon

NOTES
1. From *The Oxford Translation of Aristotle*, tr. and ed. by W.D. Ross (Oxford: Oxford University Press, 1928), "Politics," 1259a3.
2. From *Letters to Lucilius (Epistolae Morales ad Lucilium)*, tr. by Robin Campbell (Penguin Classics, 1969).
3. From *The Dialogues of Plato*, tr. by Benjamin Jowett (Oxford: Clarendon Press, 4th edition, 1953), "Eryxias," 405E.
4. From *The Oxford Translation of Aristotle*, "Rhetoric," 1387a16.
5. From *The Dialogues of Plato*, "Republic," IV, 421B.
6. From *The Oxford Translation of Aristotle*, "Politics."
7. Jean-Jacques Rousseau, *Confessions* (New York: E.P. Dutton, Everyman's Library Edition, 1911), I.
8. Ralph Waldo Emerson, *Selected Essays* (New York: Penguin, 1982), "Wealth."
9. Rousseau, *op. cit.*
10. Francis Bacon, *Essays* (Totowa, N.J.: Rowman and Littlefield University Library, 1972), "Of Seditions and Troubles."

Money and the City

David Appelbaum

COAXED BY his companion Phaedrus, Socrates walks outside the city wall. It is a day of high summer and they find an idyllic spot under a plane tree along the grassy bank of the river. The *agnos* are in full bloom, perfuming the air, while overhead cicadas drone. Phaedrus is astounded by a peculiarity of Socrates, who lives never "so much as setting foot outside the walls." What keeps Socrates confined to the city? "I'm a lover of learning," he explains, "and trees and open country won't teach me anything, whereas men in the town do." [*Phaedrus* 230d] What indispensable condition, belonging to the city alone, nourishes Socrates' search for self-knowledge?

An obvious feature of the city— Socrates' Athens—is commerce. In the *plaka* below the Acropolis, the merchants, traders, potters, smiths, and weavers ply their trades. The money-changers' hawking contrasts sharply with the still countryside. The jingle of gold and silver coins is scarcely audible. Yet it is the sound about which the urban hub revolves. City dwellers —Socrates' teachers—are caught up in an endless round of exchanging things for money. If they aid Socrates in his quest for a higher reality, *what* then is the object of their desire, money?

Another citizen of Athens, Aris-

totle, offers a clue. "All things that are exchanged," he says, "must be somehow comparable. It is for this end that money has been introduced, and it becomes in a sense an intermediate; for it measures all things." [*Nicomachean Ethics* 1133a 18]

Money is the yardstick of value in the exchange of one thing for another. It belongs to the class of great mental inventions known as measures. Mind, memory, and money all derive from the Indo-Europan root *me-* or *men-*, meaning "to measure." Measure fixes limits, degrees, amounts, and quantities. Measures of distance—the meter or mile—span the gulf between two things or places yet are not themselves things or places. They are abstractions to which we assign numerical weight. Similarly money brings things of different value together without becoming one or the other. It thereby regulates exchange. If utter difference reigns, no one can know how much of *this* to trade for how much of *that*.

Money is also power and a sign of power, a power that derives from the mind-invented measure. A measure controls—empowers—activities in a certain field. To consolidate his power, Napoleon replaced older measures with the meter as the new standard of distance. To stipulate what measure governs exchanges of value is to possess power. The Bank of England, to gain power over trade, took over the minting of currency from private banks. The power of money is amplified by the fact that it governs practical activities of central import. Commerce, giving and getting things from others, is the means to obtain the necessities of ordinary life. The more money one possesses, the greater one's power to regulate the particulars of survival, one's own and that of others. Since power attracts people's attention, money is a great motive force that extends into all areas of human life.

BESIDES MIND and measure, money has a crucial connection to memory. The Greek goddess of memory, *Mnemosyne*, became the Roman goddess *Moneta*. *Moneta* was an epithet of Juno at whose temple in Rome coinage was first struck. *Moneta* was understood to mean "The Warner," the kind of memory that we associate with admonishment. That money, the incomparable measure of exchange-value, has roots in memory and comes into our language with a warning helps explain Socrates' concern with men of money.

What is the warning that money brings? A clue can be found if we look to another momentous invention of measure, writing. Written records keep ideas in a form apparently immune to the impermanence that speech is subject to. Since we then rely on such records to learn about ourselves and the world, writing exerts a powerful influence over experience. It controls perceptions in ob-

vious and subtle ways and perpetuates a view of reality. Writing, a potent new invention in Socrates' time, provided a giant technological step forward. Yet regarding writing, Socrates warns:

If men learn this, it will implant forgetfulness in their souls; they will cease to exercise memory because they rely on that which is written, calling things to remembrance no longer from within themselves, but by means of external marks. What you have discovered is a recipe not for memory, but for reminder. [*Phaedrus* 275a]

The parallel admonishment is concealed in the history of money. However it facilitates and enlarges value-exchange, money warns us that it breeds dire forgetfulness.

What is forgotten? Money the measure (like writing) had become an unalterable condition of urban life by Socrates' day. That money is a measure, mind-invented and in itself, not real, was on the verge of being forgotten. Those measured by their relation to money relied on its power for valuation of experience. No longer were they able to remember value "from within themselves." Furthermore, ceasing to exercise memory, they grew unmindful of an important fact: that measure must derive from the recognition of lawfulness if it is not to be merely arbitrary.

THE CITY attracts Socrates with its endless circulation of money. In Athens, he is, as Phaedrus says, "the oddest of men," shunning the country in order to bear witness to the warning that money sounds. His task is to remind others of the difference between the measure and what is measured, and of that which lies beyond all measurement. To the others who locate value by means of "external marks"—coins and currency—Socrates' interrogation recollects a source of value that is absolute, unconditioned, and incorruptible.

Socrates' task in relation to money is nowhere more visible than in the matter of payment. Poor and living in squalor, he refuses payment for his teaching even though his pupils come from the monied class. His refusal, moreover, goes against custom. Men of philosophical expertise came to Athens from all over the Mediterranean to hire themselves out as tutors. For this practice and its perpetrators—whom he dubs Sophists—Socrates reserves his harshest criticism.

The critique—and the vitriol with which Socrates laces it—demands proper understanding. Clear about his own choice, Socrates does not advocate poverty as a condition for being a teacher. In itself money does not taint, though it breeds incaution and forgetfulness. Nor does he blame the Sophists in the way in which one might those unscrupulous merchants who deceive customers in order to reap profit. The Sophists err in a more fundamental direction. By accepting money for inquiry into the Self, they fail to keep the higher separate from the lower. They confound value-exchange on a single level with one between levels. Forgetting the difference between levels, they apply the measure of one to the incommensurable other.

WITHIN THE CONTEXT of teaching, what is payment? Think of the exchange between teacher and pupil. The teacher supplies the pupil with a demand for an awareness of Self, the source of real value. In return, the pupil experiences a confrontation with his or her own weakness. One is not who one thinks one is. Time and again, Socrates' interlocutors—like Phaedrus—are asked to acknowledge self-forgetting. Some embrace the demand for payment: the affirmation that one does not know the Self. A few reject it outright. Most flee to illusory self-images before they can heed the call to pay.

Contrast this exchange with that of money. Someone sells you a cord of wood. In return, you give the seller money equivalent to the accepted value of the commodity. That value derives from the wood's intrinsic worth and from the woodcutter's labor. Value in the sense of money (its cost) measures how useful, desirable, and available the thing is. When we pay money for the wood, we acknowledge the upkeep on our survival. The value of life, on this level, is comparable to the money paid to ensure its continuation.

Socrates contends that the Sophists cloud the warning that money sounds. The Sophist does not demand as payment from the student an active recognition of Self. The pupil need pay only monetarily in lieu of an inner confrontation. The Sophist leaves the encounter wealthier but without renewal of his own question. The pupil is left less vulnerable and more impervious to the truth. His belief that he understands is rooted ever more strongly in the soil of illusion.

Socrates' contention leaves us with questions. What is the place of ordinary payment in the role of a teaching? What principles guide a genuine teacher's demand for such payment?

Failure to keep Caesar's and the Lord's accounts separate spoils not only one's search but also one's money. The tainting of money, surprisingly, is no less poisonous than that of a teaching. Wary of contaminating a higher value, Socrates urges with equal caution against polluting value in the marketplace of everyday life. In Socrates' Athens, despite the fractured connection between memory and money, money is an accurate and just measure of a thing's value. Only, unmindful of how the measure and the object measured differ, one seeks money as if it possessed value in itself. One accumulates tokens of measurement, not the real things themselves.

B UT A DEEPER forgetfulness lurks. Money itself may become an inaccurate, unjust measure. Distorting the yardstick of ordinary value, money then makes it impossible to recognize true measurement. The money that once sustained exchanges of comparables now goes crazy. Cost bears no relation to the worth of an article. This degeneration of money's basic intelligence—that derives from a lawfulness recognized and acknowledged—affects attitudes. "Wealthy men," Aristotle observes,

are insolent and arrogant; their possession of wealth affects their understanding; they feel as if they had every good thing that exists; wealth becomes a sort of standard of value for everything else, and therefore they imagine there is nothing it cannot buy. . . . In a word, the type of character produced by wealth is that of a prosperous fool. [*Rhetoric* 1390b 33]

Socrates pits himself against the forces of the more total amnesia, that of unjust measure. When measure is just but is confused with the thing measured, money neither inflates nor deflates exchange value. When ordinary value stands in right relation to things, monetary value becomes transparent. The cost—of serving pots, carpets, hunting knives, goats and geese, sandals, and wine goblets—expresses a higher source of value that interpenetrates the marketplace. Just value recalls one to things' value "from within themselves," not to an arbitrary commercial value. This explains why, under conditions of justice, great care is needed in fixing cost. To err in pricing is to upset the fragile correspondence between the horizontal exchange and the vertical one. It is to cut the marketplace off from the divine guidance available to value-exchange.

How does one—merchant or customer—come to know just measure? With money as universal as it is, the question, Socrates discovers, recalls us to the fundamental human condition. Fair pricing points back to the justness of the value-maker. Not a business course but training in basic awareness is a prime ingredient in a person's perception of justice. Greed, miserliness, fraud, and exploitation stem, for Socrates, from an inattention to the question of inner harmony. When life is lived disharmoniously, thought defrauds feeling and feeling exploits the appetites. Lacking the means to restore the dynamic balance, one suffers the ignorance of one's own just measure.

Justice belongs to the person in whom each part performs the function proper to it, without being overruled by another part. How can the teacher transmit this disciplined training to people who have been deformed by theft, cheating, deception, and other unjust practices of value-exchange? Nothing is possible, Socrates sees, until a man is stopped in his tracks. Recognition of the impoverished relation to Self—amidst the tangible wealth of the world—can alone awaken one to the need for instruction. But for the prosperous fool who dreams his empty dreams of riches? "And if to a man in this state of mind," Socrates asks,

someone gently comes and tells him what is the truth, that he has no sense and sorely needs it, and that the only way to get it is to work like a slave to win it, do you think it will be easy for him to lend an ear to the quiet voice in the midst of and in spite of these evil surroundings? [*Republic* 494d]

Is that quiet voice not the warning that money issues? Around the marketplace Socrates prowls, listening. He has accepted the aim of bringing others to heed its warning and attend to the Self—the allegory of the Cave imparts this lesson. There he awaits the moment to deliver others from self-forgetfulness into listening. The deforming force that trades the measure for the thing itself is, however, too potent. The many bright young men of the city—Pheadrus, Euthyphro, Charmides, Euthydemus—incline toward wealth rather than well-being. Their inner aspects already manifest the imbalance symptomatic of unjust dealings. Yet as he tries to carry out his task, Socrates is deeply struck by the obvious. The current conditions of commerce militate against others' being available to his teaching. His words fall on deaf ears. The deeper forgetfulness has begun to set in.

AS HE WALKS through Athens' many bazaars, Socrates has an insight. Valuing money correctly, the city itself plays an absolutely essential part. It stands midway between man and the cosmos. Man as a partner of just measure depends on the city in the same way a linguist depends on a dictionary. The city is the book of justice. If the text is corrupt and man corrupted by urban commerce, he also can be remade by it. *The Republic* is the blueprint for urban renewal.

As an intermediate, the city (like money) makes unlike things (man and cosmos) comparable. Although rarely so considered, the city also is a measure. It gives outwardly a measure of human accomplishment in its buildings, avenues, and plazas. More subtly, the city provides a measure of inner life through its art, sacred spaces, and places of learning. The two aspects are meant to fit together in the way that heart informs deed. When inner and outer are related, measure allows things of unlike scale—the human and the celestial—to be encompassed within the same locale, the city walls. The passage of energy between man and cosmos is regulated, blended, and made concordant. Each "partner" flourishes.

When the inner aspect of measure is neglected, we have the wasteland. Whether a fascistic model of order and prosperity or a decaying urban ghetto filled with the debris of human life, such a city provides a flawed measure. This is T.S. Eliot's unreal city ("Jerusalem Athens Alexandria/Vienna London"), decadent, serving no meaningful human end. All measures contained therein—from language to record-keeping—are distorted and unjust. When the exchange between humanity and what lies above is unregulated, the needs of both man and cosmos go unanswered. Human relations of all kinds are bent. So too is commerce, the lifeblood of the city. The exchange of things of value is drastically altered since money, like all measure, is crooked.

Socrates' plan to restore justice is striking. Measure, if just, is law-conformable. To embody justice, the city in all its aspects needs again to conform to law. It needs to cease to be ruled by self-will, individual or collective, the rule of an outlaw. Such tyranny results from an internal imbalance in which—as in a person—the thinking, feeling, and appetitive parts of the metropolis vie for each other's power. When by contrast all constituents act in accordance with their specific excellences or virtues, the city as a whole finds itself in agreement with the laws governing the kind of thing that it is. This wholeness alone guarantees that the city as measure is free from distortion.

In the marketplace, Socrates again hears the sound of coins. Faint but sharp, it is the voice of *Moneta's* warning. She asks, Does my measure conform to law? Do you keep the rigor of mind needed to preserve it? ●

EPICYCLE

Two Mullah Judgments/Middle Eastern

SANIE HOLLAND

The Sack of Gold

ONCE THE MULLAH, nagged by his wife as usual for being out of work, announced, "Taking a job is beneath my dignity. I am in the service of Allah."

"If that is so," replied his wife, "please be so kind as to petition Allah for back wages."

This struck the Mullah as being an excellent idea, and he immediately retired to his garden, got down on his knees, and began to pray in a loud voice.

"Allah, hear me!" he cried. "You know I have served you well. For all my past services which have gone unrewarded, I beg you to grant me a hundred pieces of gold!" On and on he continued in similar fashion, at the top of his voice. His neighbor, overhearing him, thought to play a trick on the Mullah. Taking a sack, he filled it with a hundred pieces of gold and flung it over the wall into the Mullah's garden. The Mullah, astounded by the efficacy of his prayers, thanked Allah and took the money inside.

"Allah has rewarded me," he announced to his wife, and calmly showed her the sack of gold. She was no less astounded than he had been. Thereafter, the two of them began to live somewhat less penuriously. Their neighbor, annoyed at how blithely his joke had been taken at face value, called on the Mullah to ask for his money back.

"You heard my prayers, as did Allah, and now you are pretending it was you who answered them. Be off with you!"

"If you will not return the money," said the neighbor, "I will take you before the Cadi."

The Mullah thought for a moment. "If you are determined to do this," he said, "then let us get it over with at once. However, I have not clothes as suitable as yours, and only a donkey to ride on. I fear the case would be prejudiced in your favor on appearances alone. That would be unjust, and give you as little satisfaction as it would me. Lend me a robe, if you please, and a horse. That will make things appear equitable." The neighbor grudgingly agreed, and the two of them set off.

When they went before the Cadi, the neighbor's case was heard first. When the Mullah was asked to defend himself, he replied that his defense was insanity: his neighbor was insane.

"How so?" the Cadi asked. "What is the evidence?"

"If you ask him," replied the Mullah, "he will claim that everything is his: not only the money, but the horse I rode on and even the clothes on my back."

"But they are indeed mine!" cried the neighbor.

The case was dismissed.

Payment in Kind

FROM TIME TO TIME, whenever the Cadi who governed the village and gave judgments in the court was absent, the role was given to Nasr Eddin Hodja. It was under just such circumstances that one day there was presented to the court an unusual and difficult case. A local innkeeper brought suit against a poor student in these terms:

"He has lingered outside my restaurant, effendi, and has thus stolen from me."

"And what has he stolen?" inquired the Hodja.

"The good smell of my good food," replied the innkeeper. "Unwilling to pay for the food itself, he has lingered around the door of my kitchen daily, and availed himself of what was not his: the aromas of my cooking. Thus I work and slave, and this scoundrel takes advantage of my labors and will not pay."

"Is this true, young man?" the Hodja demanded.

"It is, effendi. I am a poor student, scarcely able to pay for my room and my books. I live on scraps which I beg wherever I can. But the wonderful smells from the inn I could not resist, and so daily I hang about the kitchen and imbibe those odors, and thus imagine that I am eating those very delicacies."

"And have you now any money on you?" asked the Hodja.

"Only a few coppers, effendi," was the reply.

"Hand them over."

As the poor student passed his last coins to the Hodja, the innkeeper smiled with satisfaction. The Hodja turned to him then and said, "Innkeeper, close your eyes, and listen well to my judgment."

Puzzled, the innkeeper did so. And then, with his eyes tightly shut, he heard the student's coins being jangled in the Hodja's hand.

"Do you hear, innkeeper?" asked the Hodja.

"I hear, wise effendi," replied the innkeeper.

"Good! The sound of the coins has paid for the smell of the food," replied the Hodja, as he returned the coins to the student.

— *Retold by Paul Jordan-Smith*

Your Money or Your Life

Joseph Kulin

MY FATHER, WHO iden-
tified with the par-
simonious comedian, Jack Benny,
loved to repeat the skit about a mugger
who accosts Benny and says, "Your
money or your life!" After an appre-
ciable silence, the mugger says,
"Well?" and Benny replies, "Don't
rush me, I'm thinking, I'm thinking!"

Once again, the U.S. is entering
a major recession, and as thousands
are laid off, banks fail, homeless wan-
der the streets, bankruptcies prolifer-
ate, the dollar hits new lows and the
national debt new highs, questions
about money assault us and take on
added urgency. What is this thing that
so affects our lives, our thinking, our
valuation of ourselves and our society?
From what past has money emerged,
and to what future is our current relationship to money leading us?

Money originally was a storehouse of power. In its most primal
sense, what eventually became

"money" must have been regarded as
magical, something able to translate
differences between commodities or
services and to reconcile relationships
on different levels. Money was a symbol, a metaphor for the Divine, the
Creator, having the power to create,
to do. Utilized ceremonially, it had
the ability to attract the attention of
spirits or gods. Cattle, often a measure of wealth, were commonly used
in sacrifices, and in the Orient, paper
money tied to trees or burnt is still
used for this purpose. In this sense,
money could bestow prestige, settle
psychic and material debts, placate enemies, and sanction transactions, all
by virtue of its power to represent
higher principles or authority.

In its earliest origins, therefore,
money was some commodity that had
intrinsic value, held in common esteem by a community, and the various
forms it took, such as salt, cattle, fish,
cowrie shells, jewelry, glass beads,

feathers, gold, or even human skulls, depended very much on local conditions. But that simple, direct relationship to money has greatly changed over the ages and we see how as our world has grown in sophistication and complexity, so too has money assumed different garb and functions reflecting the various needs and ambitions of the society it served.

In traditional societies, great importance was put upon the maintenance of balance and order within the community as well as with the surrounding environment. What passed as currency in ancient times was probably controlled by shamans or priests, who didn't actually own the money but were responsible for the proper circulation and use of it in the community. (To this day, Native American shamans and medicine men are among the poorest people of their tribe, and intentionally so.) In cultures where commodities were in short supply, the recycling and repeated use of money, no doubt, enhanced its value, added to the esteem of the community, and strengthened social bonds. In cultures where there was great abundance of natural wealth, traditional wisdom prescribed potlatches (community giveaways) or destruction of surplus food or goods in order to maintain harmony and prevent class differences from arising. In the latter case, great moral virtue was attributed to those who disdained the personal possession of large stores of fish, rice, copper shields, or other emblems of wealth, and who demonstrated that nonattachment by publicly destroying or giving away what they had.

Having somehow lost its harmonious relationship with the natural world, the Web of Life, modern society has become increasingly unbalanced, like a top spinning out of control. In a thought-provoking essay, "Basic Call to Consciousness,"[1] written by tribal elders of the Iroquois Nation, we are challenged to reconsider what Western mentality has assumed to be some of civilization's greatest early achievements. In particular, we are asked to ponder the harmful effects that resulted from the departure from a hunting and gathering way of life based on a respectful, balanced relationship with Nature, for a life of increasing manipulation and domination of animals, water, natural resources, and people.

Similarly, the introduction of coinage, commonly perceived as one of humanity's great intellectual innovations, appears to have become a contributing factor in undermining a more spiritual relationship to the world. The issuance and regulation of currency by governments has served as an effective means whereby exploitative groups have been able to extend and consolidate their power and control.

*The nominal value was
validated and backed
by faith and trust
in the king.*

MONEY WAS FIRST minted
in the Mediterranean re-
gion when trade between nations was
burgeoning and conditions ripe for the
introduction of a more convenient medium of exchange, something more
universal, portable, and durable. State
coinage is said to have originated in
the kingdom of Lydia in the seventh
century B.C. and coincided with the
first recorded "retail shops." The es-
tablishment of minting enabled rulers
to issue coins whose nominal value
exceeded their value as metals. The
nominal value was validated and
backed by faith and trust in the king
or rulers issuing the coinage. But
when power passed from sacred to
temporal rulers, the safeguards of the
past were left behind. The new rulers
were not concerned with maintaining
balance and stability within the community but with causing an imbalance
by gathering more wealth and power
for themselves. These rulers no longer
saw themselves as caretakers of
wealth, but as owners.

Unfortunately, the history of
money is often a tale of trust mis-
placed or betrayed. In medieval times,
according to an entry by David Car-

penter in *The Encyclopedia of Religion*,
the gap between the sacred and the
profane—money as mere coinage—
grew even greater:

Money was increasingly represented as de-
monic. Feelings of awe before the numi-
nous qualities of gold and silver were
transformed into feelings of disgust for
gold and silver coins, which were increasingly compared to excrement.[2]

Europe of the Renaissance was
marked by the frenzy for money, and
gold was the new mark of wealth,
more useful than land because it could
buy anything. Such ambassadors to
the New World as Christopher Columbus and the conquistadors in Mexico and Peru were infected with the
same avarice, an obsession with gold,
with the result that the native peoples
of the Americas were extensively ex-
ploited and slaughtered.

But Europe was not ultimately
any better off; in fact, the imbalance
within its society grew only greater.
Hans Koning writes:

For all the gold and silver stolen and
shipped to Spain did not make the Spanish
people richer. It gave their kings an edge
in the balance of power for a time, a
chance to hire more mercenary soldiers for
their wars. They ended up losing those
wars anyway, and all that was left was a
deadly inflation, a starving population, the
rich, richer, the poor, poorer, and a ruined
peasant class.[3]

The tone for the following five
hundred years was being set.

The frenzy in the early capitalist states of
Europe was for gold, for slaves, for products of the soil, to pay the bondholders and
stockholders of the expeditions, to finance
the monarchical bureaucracies rising in
Western Europe, to spur the growth of the
new money economy rising out of feudal-
ism, to participate in what Karl Marx
would later call "the primitive accumulation of capital." These were the violent beginnings of an intricate system of technology, business, politics, and culture that
would dominate the world for the next five
centuries.[4]

In order to better facilitate these intricate new systems, money has taken on progressively abstract, representational forms. Paper currency which became popular about 300 years ago was initially convertible into gold or silver. We then had paper money backed by "full faith and credit" of governments, commercial notes or bank checks, and most recently, electronic credit card systems. Thus, the nature of money has been transformed over time from commodities of intrinsic value, to hard coinage, to printed paper representations, to electronic data. The most telling indicator of personal wealth these days is provided in the form of computer printouts, and the best way to make instant profits is by getting "inside information." However, the further we depart from the original source and sense of money, the greater the likelihood that we may be misleading ourselves.

Money is not really wealth: it is a device for exchanging and measuring wealth. Any increase or decrease in the quantity of currency in a country does not necessarily constitute an increase or decrease in the country's wealth. Its universal acceptance and liquidity seem to be based on a kind of hypnotic circular reasoning: money is so acceptable just because it is so readily accepted. We are willing to accept money in exchange for things of value because we have no doubt that we will be able to pass the money on to others just as easily. Apart from its communal context, money becomes meaningless; it sustains the illusion of value only as long as the social bond holds.

WHAT HAPPENS WHEN the consensus, the communal trance, is broken and the bubble bursts? There is the famous example of tulipomania in Holland in 1637, when tulips became the subject of extraordinary speculation and prices for single bulbs rose to enormous sums. Thousands of people became bankrupt once the illusion became evident. And of the great stock market crash of October, 1929, it has been written:

The country's real wealth had not been diminished by so much as a single automobile tire or one railroad engine, but its paper wealth as measured in stock prices had been cut almost in half. . . . Some $35,000,000,000 had vanished from the economy. Where had the wealth gone? Hundreds of thousands of peoples who had considered themselves rich now considered themselves poor. Stock certificates that only yesterday had been the keys to wealth, power, and position were suddenly pieces of printed paper without worth. No amount of cheerleading could reverse the magic and make them valuable once more.[5]

Money can act like a narcotic giving us "highs" when we have it and leaving us depressed when we don't. The infusion of money stimulates action, and accelerates the current of life. It greases the wheels of trade, production, exploration, and increases the complexity and tempo of our lives. Quantity becomes more emphasized, more highly regarded than quality of life. Time becomes an increasingly limited resource, for "time is money," and we prefer the right hand side of the equation.

Money is wrongfully debased when reduced to purely economic terms, but what means can we take to redeem our relationship to it? First, we need to see clearly its effect on us, and the extent to which we are identified with it and the allures it offers. We need to see how much we desire it and are never satisfied with what we have; how the pursuit of money is responsible for our lack of time; how we measure our prestige, our sense of self-esteem by our wealth and possessions; how we grudgingly admire the fortunes of the Trumps, Milkens and Boeskys while they are temporarily on

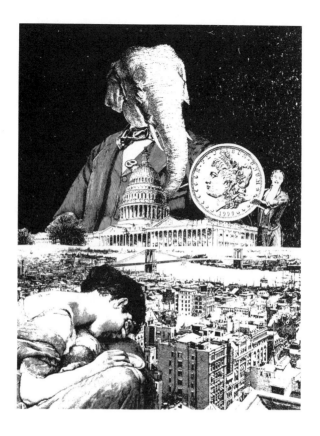

top. Until we wake up to the fact of society's and our own addiction, we will remain under money's sway.

My own views concerning money have been greatly influenced by my early upbringing. My father came to this country in 1910. He worked as a professional tailor and was able to acquire, after a few decades, a house and some savings. His world turned upside down in the early 1930s when the bank failures of the Great Depression resulted in the loss of his house and all his savings, and he was unable to find work as a tailor. In order never to be caught in the same situation again, he took a secure job as a laborer in a restaurant and worked for another twenty years before retirement, never earning more than minimum wages while raising a family of five.

Needless to say we were sparing in our wants, but we could afford the basics. My father showed us how to make do with little. There were no extras, no frills. We lived within our means. He didn't look for opportunities to make more money; he turned his back on monetary ambition. He led a simple, disciplined life and always had time to play ball with me and to take walks in Prospect Park on weekends. He was very careful with his money, never bought anything on credit, but was not stingy—I regularly received an allowance, with which I learned to operate. He knew how to get value for his expenditures, and nothing went to waste. He taught me to respect money, but not to worship it. He lived to the age of ninety-three, after twenty-eight years of enjoyed retirement.

For our own betterment and for others, those spiritually oriented summon us to revalue our values, to consciously scrutinize our attitudes to money. Helen M. Luke advises us to elevate the value of our exchanges by honoring equally both the paying and earning within our daily transactions. She writes:

When there is true exchange between employer and employed it is a beautiful thing —the master respecting the skills and service of the man, the man respecting the master's knowledge and willingness to take responsibility and risk. It can still today be a relationship full of that dignity and true equality so easily lost the minute everyone starts fighting for their "rights" instead of for the fairness of the exchange. As soon as earning and paying become bargaining, the issues turn ugly; the love of money in both employers and employees takes over as an end in itself.[6]

W E ARE SLOWLY awakening to the fact that we are approaching the end of the earth's natural resources. This natural wealth, which was once in the safekeeping of the indigenous peoples of the world, has now nearly all been stolen, destroyed, consumed, or polluted. Perhaps we could still learn to be better custodians, if we would realize our responsibility to see that this wealth is employed for the benefit of all life, now and unto seven generations, as Iroquois tradition holds.

Why has Joseph Campbell's summons to "Follow your bliss!" struck such a responsive chord among so many? It is a liberating call to pay more heed to our deeper nature and less to the obviously dying premises of our society, the collective psychosis, that suggests to us that our lives should be spent getting, accumulating, and consuming outer things. There is now a growing recognition that the great imbalances our society has caused must somehow be rectified before it is too late. As E.F. Schumacher has written:

The chances of mitigating the rate of resource depletion or of bringing harmony into the relationship between those in possession of wealth and power and those without it is non-existent as long as there is no idea anywhere of enough being good and more-than-enough being evil.[7]

The pursuit of wealth, power, esteem, leads us outside, away from ourselves. We need to appreciate once more the wisdom of the golden mean, wise moderation, the avoidance of extremes. Money was once a means whereby balance could be maintained or restored within a community. We need to find again that function for it. But, only by finding a center of balance and order in oneself can a new harmony, both inner and outer, be reestablished. The experience of true fulfillment and richness is found within: the outer world may then be helped back to health by virtue of the manifestation of this new inner order.   •

NOTES

1. From *Basic Call to Consciousness* (Mohawk Nation: Akwesasne Notes, 1978), pp. 71-74.
2. Mircea Eliade, ed., *The Encyclopedia of Religion* (New York: Macmillan Publishing Company, 1987), Vol 10, "Money."
3. Hans Koning, *Columbus: His Enterprise* (New York: Monthly Review Press, 1982).
4. Howard Zinn, *A People's History of the United States* (San Francisco: Harper & Row, 1981).
5. John Rublowsky, *After the Crash* (New York: Crowell-Collier, 1970), pp. 68-69.
6. Helen M. Luke, *Woman, Earth and Spirit* (New York: Crossroad Publishing Co., 1989), p. 88.
7. E. F. Schumacher, *Small Is Beautiful* (New York: Harper & Row, 1973), p. 297.

Making Real Gold

Titus Burckhardt

ALCHEMY HAS existed since at least the middle of the first millennium before Christ, and probably since prehistoric times. To the question: why should alchemy have existed for millennia in such widely separated civilizations as those of the Near East and Far East, the reply of most historians would be that man has repeatedly fallen to the temptation to get rich quickly by seeking to make gold and silver from the common metals, until the empirical chemistry of the eighteenth century finally proved to him that metals could not be changed into one another. In actual fact, however, the truth is quite different, and, in part at least, exactly the opposite.

Gold and silver were already sacred metals, even before they became the measure of all commercial transactions. They are the earthly reflections of sun and moon, and thus also of all realities of spirit and soul which are related to the heavenly pair. Until well into the Middle Ages the relative values of the two noble metals were determined by the relationship of the rotation times of the two heavenly bodies. Likewise, the oldest coins usually bear pictures or signs having some relation to the sun or its yearly rotation. For men of pre-rationalistic times

the relationship between the noble metals and the two great luminaries was obvious, and a whole world of mechanistic notions and prejudices was necessary to obscure the self-evident reality of this relationship and make it look like a mere aesthetic accident.

One must not confuse a symbol with a mere allegory, nor try to see in it the expression of some misty and irrational collective instinct. True symbolism depends on the fact that things, which may differ from one another in time, space, material nature, and many other limitative characteristics, can possess and exhibit the same essential quality. They thus appear as various reflections, manifestations, or productions of the same reality— which in itself is independent of time and space. It is thus not quite right to say that gold represents the sun, and silver the moon; rather is it the case that the two noble metals and the two luminaries are both symbols of the same two cosmic or divine realities.[1]

THE MAGIC OF gold thus springs from its sacred nature, or qualitative perfection, and only secondarily from its economic value. In view of the sacred nature of gold and silver, the obtaining of these two metals could only be a priestly ac-

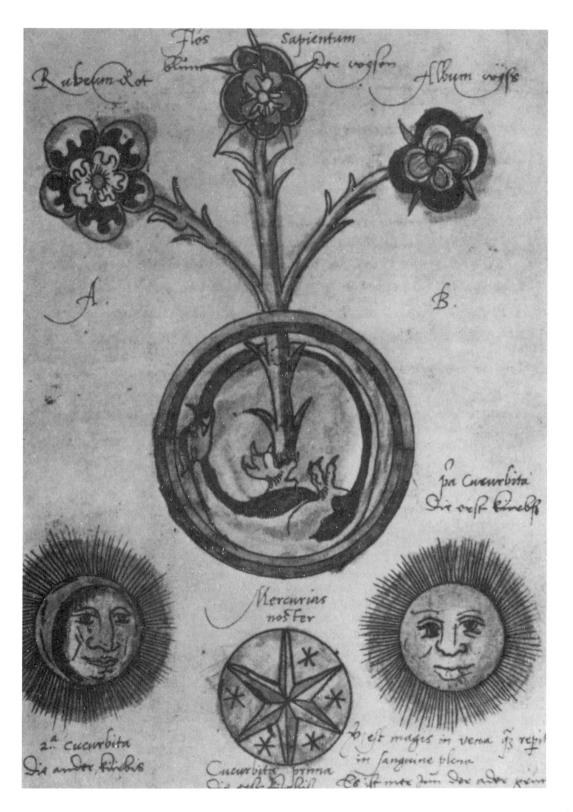

Rubeum Rot   Flos   Sapientum   Album wiß
                blum   der weisen

A.   B.

pa Cucurbita

Mercurius
noster

2ª cucurbita   est magis in vena que reple-
die ander kürbß   Cucurbita prima   in sanguine plena

tivity, just as the minting of gold and silver coins was originally the prerogative of certain holy places only. In keeping with this is the fact that the metallurgical procedures relating to gold and silver, which have been preserved in some so-called primitive societies from prehistoric times, reveal abundant signs of their sacerdotal origin.[2] In "archaic" cultures, unacquainted as yet with the dichotomy of "spiritual" and "practical," and seeing everything in relation to the inner unity of man and the cosmos, the preparation of ores is always carried out as a sacred procedure. As a rule it is the prerogative of a priestly caste, called to this activity by divine command. Where this is not so, the smelter or smith, as an unauthorized interloper into the sacred order of nature, falls under suspicion of engaging in black magic.[3] What appears to modern man as superstition—and what, in part, only survives as a superstition—is in truth an inkling of the profound relationship between the natural order and the human soul. "Primitive" man is well aware that the production of ores from the "womb" of the earth and their violent purification by fire is somehow sinister, and fraught with dangerous possibilities, even if he does not have all the proofs with which the history of the metallurgical age has so abundantly provided us. For "archaic" humanity—which does not artificially separate matter from spirit—the arrival of metallurgy was not simply an "invention," but rather a "revelation," for only a divine command could authorize mankind in respect of such an activity. From the very beginning, however, this revelation was very much a "two-sided" one;[4] it demands an especial prudence on the part of him to whom it has been given. Just as the outward work of the metallurgist with ore and fire has something violent about it, so also the influences which

bear back on the spirit and the soul—and which are inescapable in this calling—must be of a dangerous and two-sided nature. In particular, the extraction of the noble metals from impure ores by means of solvent and purifying agents such as mercury and antimony and in conjunction with fire, is inevitably carried out against the resistance of the darksome and chaotic forces of nature, just as the achievement of "inward silver" or "inward gold"—in their immutable purity and luminosity—demands the conquest of all the dark and irrational impulses of the soul.

That there is an inward gold, or rather, that gold has an inward as well as an outward reality, was only logical for the contemplative way of looking at things, which spontaneously recognized the same "essence" in both gold and the sun. It is here, and nowhere else, that the root of alchemy lies. Alchemy traces its descent back to a priestly art of the ancient Egyptians; the alchemical tradition which spread all over Europe and the Near East, and which perhaps even influenced Indian alchemy, recognizes as its founder Hermes Trismegistos, the "thrice-great Hermes," who is identifiable with the ancient Egyptian God Thoth, the God who presides over all priestly arts and sciences, rather like Ganesha in Hinduism.

IN FAVOR OF an Egyptian origin of Near-Eastern and Western alchemy is the fact that a whole series of artisanal procedures, related to alchemy and furnishing it with many of its symbolic expressions, crop up as a coherent group, from late Egyptian times onwards, finally making an appearance in medieval prescription books. This corpus of procedures contains some elements clearly derivable from Egypt. Among these procedures, apart from the working of metal and the preparation of dyes, is the production of artificial precious stones and colored glass, an art which blossomed nowhere more than in Egypt. Moreover, the whole of the Egyptian art of metals and minerals, in its efforts to extract the secret and the precious essence from an earthly "substance," shows an obvious spiritual relationship with alchemy.

Late Egyptian Alexandria was doubtless the melting-pot in which alchemy, along with other cosmological arts and sciences, received the form in which it is now known to us, without thereby being altered in any essential respect. It may well have been at that time also that alchemy acquired certain motifs from Greek and Asiatic mythologies. This must not be regarded as in any way an artificial happening. The growth of a genuine tradition resembles that of a crystal, which attracts homologous particles to itself, incorporating them according to its own laws of unity.

From this time onwards one can observe two currents in alchemy. One is predominantly artisanal in nature; the symbolism of an "inward work" appears here as something superadded to a professional activity and is only mentioned occasionally and incidentally. The other makes use of metallurgical processes exclusively as analogies, so that one may even question whether they were ever employed "outwardly." This has caused some to make a distinction between artisanal alchemy—believed to be the older—and a so-called mystical alchemy which is supposed to have developed later. In reality, however, it is a case of two aspects of one and the same tradition, of which the symbolical aspect is undoubtedly the more "archaic."

Meanwhile the symbolism of alchemy, as a result of its gradual incorporation into late classical and Semitic thought, had developed into a variegated multiplicity. Nevertheless certain fundamental traits proper to alchemy as an "art" remained as its infallible token throughout the centuries: above all the definite plan of the "alchemical work," the individual phases of which are characterized by certain "symbolical" processes which cannot always be carried out in practice.

IN MY BOOK on the principles and methods of sacred art[5] I more than once had occasion to refer to alchemy, by way of comparison, when considering artistic creation as it appears within a sacred tradition, not from the point of view of its outward aesthetic aspect, but as an inward process whose goal is the ripening, "transmutation," or rebirth of the soul of the artist himself. Alchemy was called the "royal art" (*ars regia*) by its masters, and, with its image of the transmutation of base metals into the noble metals gold and silver, serves as a highly evocative symbol of the inward process referred to. In fact alchemy may be called the art of the transmutations of the soul. In saying this I am not seeking to deny that alchemists also knew and practiced metallurgical procedures such as the purification and alloying of metals; their real work, however, for which all these procedures were merely the outward supports or "operational" symbols, was the transmutation of the soul. The testimony of the alchemists on this point is unanimous. For example, in *The Book of Seven Chapters*, which has been attributed to Hermes Trismegistos, the father of Near-Eastern and Western alchemy, we read: "See, I have opened unto you what was hid: The [alchemical] work is with you and amongst you; in that it is to be found within you and is enduring; you will always have it present, wherever you are, on land or on sea . . ."[6] And in the famous dialogue between the Arab king Khalid and the sage Morienus (or Marianus) it is told how the king asked the sage where one could find the thing with which one could accomplish the Hermetic work. At this Morienus was silent, and it was only after much hesitation that he answered: "O King, I declare the truth to you, that God in His mercy has created this extraordinary thing in yourself; wherever you may be, it is always in you, and cannot be separated from you . . ."7

From all this it will be seen that the difference between alchemy and any other sacred art is that in alchemy mastery is attained, not visibly on the outward artisanal plane as in architecture and painting, but only inwardly; for the transmutation of lead into gold which constitutes the alchemical work far exceeds the possibilities of artisanal skill. The miraculousness of this process, effecting a "leap" which, according to the alchemists, nature by herself can only accomplish in an unforeseeably long time, highlights the difference between corporeal possibilities and those of the soul. While a mineral substance, whose solutions, crystallizations, smeltings, and burnings can reflect up to a point the changes within the soul, must remain confined within definite limits, the soul, for its part, can overcome the corresponding "psychic" limits, thanks to its meeting with the Spirit, which is bound by no form. Lead represents the chaotic, "heavy," and sick condition of metal or of the inward man, while gold— "congealed light" and "earthly sun"— expresses the perfection of both metallic and human existence. According to the alchemists' way of looking at things, gold is the real goal of the metallic nature; all other metals are either preparatory steps or experiments to that end. Gold alone possesses in itself a harmonious equilibrium of all metallic properties, and therefore also pos-

sesses durability. "Copper is restless until it becomes gold," said Meister Eckhart, referring in reality to the soul, which longs for its own eternal being. Thus, in contradistinction from the usual reproach against them, the alchemists did not seek, by means of secretly conserved formulas in which only they believed, to make gold from ordinary metals. Whoever really wished to attempt this belonged to the so-called "charcoal burners" who, without any connection with the living alchemical tradition, and purely on the basis of a study of the texts which they could only understand in a literal sense, sought to achieve the "great work."

With its "impersonal" way of looking at the world of the soul, alchemy stands in closer relation to the "way of knowledge" (gnosis) than to the "way of love." For it is the prerogative of gnosis—in the true and not the heretical sense of the expression— to regard the "I"-bound soul "objectively," instead of merely experiencing it subjectively. This is why it was a mysticism founded on the "way of knowledge" that on occasion used alchemical modes of expression, if in fact it did not actually assimilate the forms of alchemy with the degrees and modes of its own "way." The expression "mysticism" comes from "secret" or "to withdraw" (Greek *myein*); the essence of mysticism eludes a merely rational interpretation, and the same holds good in the case of alchemy.

ANOTHER REASON why alchemical doctrine hides itself in riddles is because it is not meant for everyone. The "royal art" presupposes a more than ordinary understanding, and also a certain cast of soul, failing which its practice may involve no small dangers for the soul. "Is it not recognized," writes Artephius, a famous alchemist of the

Middle Ages,[8] "that ours is a cabbalistic art? By this I mean that it is passed on orally, and is full of secrets. But you, poor deluded fellow, are you so simple as to believe that we would clearly and openly teach the greatest and most important of all secrets, with the result that you would take our words literally?"

One may well be surprised that, in spite of such warnings, of which many more examples could be furnished, many people—especially in the seventeenth and eighteenth centuries—believed that by diligent study of the alchemical texts they would be able to find the means of making gold. It is true that alchemist authors often imply that they preserve the secret of alchemy only to prevent anyone unworthy acquiring a dangerous power. They thus made use of an unavoidable misunderstanding to keep unqualified persons at a distance. Yet they never spoke of the seemingly material aims of their art without also mentioning the truth in the same breath. Whoever was motivated by worldly passion would automatically fail to grasp the essential in any explanation. Thus in the *Hermetic Triumph* it is written: "The philosophers' stone (with which one can turn base metals into gold) grants long life and freedom from disease to the one who possesses it, and in its power brings more gold and silver than all the mightiest conquerors have between them. In addition, this treasure has an advantage above all others in this life, namely, that whoever enjoys it will be perfectly happy—the very sight of it making him happy—and will never be assailed by the fear of losing it."[9] The first sentence appears to confirm the outward interpretation of alchemy, whereas the second indicates, as clearly as is deemed desirable, that the possession in question here is inward and spiritual.

That is the spiritual threshold which the alchemist has to cross. The ethical threshold, as we have seen, is the temptation to pursue the alchemical art only on account of gold. Alchemists constantly insist that the greatest obstacle to their work is covetousness. This vice is for their art what pride is for the "way of love" and what self-deception is for the "way of knowledge." Here covetousness is simply another name for egoism, for attachment to one's own limited ego in thrall to passion. On the other hand, the requirement that the pupil of Hermes must only seek to transmute elements in order to help the poor—or nature herself—in need, recalls the Buddhist vow to seek the highest enlightenment only with a view to the salvation of all creatures. Compassion alone delivers us from the artfulness of the ego, which in its every action seeks only to mirror itself. •

NOTES

1. An excellent description of what a symbol means is to be found in the ethnological study by E.E. Evans-Pritchard, *Nuer Religion*, chapter entitled *The Problem of Symbols* (Oxford: Clarendon Press, 1956).
2. See: Mircea Eliade, *Forgerons et alchimistes*, collection "Homo sapiens," Paris, 1956.
3. *Ibid.*
4. "We revealed iron, wherein is evil power and many uses for mankind . . ." (Koran, LVII, 25).
5. *Sacred Art of East and West* (London: Perennial Books, 1967).
6. *Bibliothèque des philosophes chimiques* (Paris: G. Salmon, 1741).
7. *Ibid.*, II. The dialogue of the Arab king Khalid with the monk Morienus or Marianus was probably the first alchemical treatise to be translated from Arabic into Latin.
8. *Ibid.*
9. *Ibid.*

Excerpted from Titus Burckhardt, *Alchemy: Science of the Cosmos, Science of the Soul,* translated by William Stoddart, (Rockbury, Mass. and Shaftesbury, England: Element Books, 1986). Reprinted by permission of Element Books, Ltd.

EPICYCLE

The Spirit in the Bottle/German

THERE WAS ONCE a poor wood-cutter who toiled from early morning till late at night. When at last he had laid by some money he said to his boy: "You are my only child, I will spend the money which I have earned with the sweat of my brow on your education; if you learn some honest trade you can support me in my old age, when my limbs have grown stiff and I am obliged to stay at home." Then the boy went to a High School and learned diligently so that his masters praised him, and he remained there a long time. When he had worked through two classes, but was still not yet perfect in everything, the little pittance which the father had earned was all spent, and the boy was obliged to return home to him. "Ah," said the father, sorrowfully, "I can give you no more, and in these hard times I cannot earn a farthing more than will suffice for our daily bread." "Dear father," answered the son, "don't trouble yourself about it, if it is God's will, it will turn to my advantage. I shall soon accustom myself to it." When the father wanted to go into the forest to earn money by helping to chop and stack wood, the son said: "I will go with you and help you." "Nay, my son," said the father, "that would be hard for you; you are not accustomed to rough work, and will not be able to bear it. Besides, I have only one axe and no money left wherewith to buy another." "Just go to the neighbor," answered the son, "he will lend you his axe until I have earned one for myself."

The father then borrowed an axe of the neighbor, and next morning at break of day they went out into the forest together. The son helped his father and was quite merry and brisk about it. But when the sun was right over their heads, the father said: "We will rest, and have our dinner, and then we shall work twice as well." The son took his bread in his hands, and said: "Just you rest, father, I am not tired; I will walk up and down a little in the forest, and look for birds' nests." "Oh, you fool," said the father, "why should you want to run about there? Afterwards you will be tired, and no longer able to raise your arm; stay here, and sit down beside me.

THE SON, however, went into the forest, ate his bread, was very merry and peered in among the green branches to see if he could discover a bird's nest anywhere. So he walked to and fro until at last he came to a great dangerous-looking oak, which certainly was already many hundred years old, and which five men could not have spanned. He stood still and looked at it, and thought: "Many a bird must have built its nest in that." Then all at once it seemed to him that he heard a voice. He listened and became aware that someone was crying in a very smothered voice: "Let me out, let me out!" He looked around, but could discover nothing; then he fancied that the voice came out of the ground. So he cried: "Where are you?" The voice answered: "I am down here amongst the roots of the oak-tree. Let me out! Let me out!" The schoolboy began to loosen the earth under the tree, and search among the roots, until at last he found a glass bottle in a little hollow. He lifted it up and held it against the light, and then saw a creature shaped like a frog, springing up and down in it. "Let me out! Let me out!" it cried anew, and the boy, thinking no evil, drew the cork out of the bottle. Immediately a spirit ascended from it, and began to grow, and grew so fast that in a very few moments he stood before the boy, a terrible fellow as big as half the tree. "Do you know," he cried in an awful voice, "what your reward is for having let me out?" "No," replied the boy fearlessly, "how should I know that?" "Then I will tell you," cried the spirit; "I must strangle you for it." "You should have told me that sooner," said the boy, "for I should then have left you shut up, but my head shall stand fast for all you can do; more persons than one must be consulted about that." "More persons here, more persons there," said the spirit. "You shall have the reward you have earned. Do you think that I was shut up there for such a long time as a favor? No, it was a punishment for me. I am the mighty Mercurius. Whoso releases me, him must I strangle." "Slowly," answered the boy, "not so fast. I must first know that you really were shut up in that little bottle, and that you are the right spirit. If, indeed, you can get in again, I will believe, and then you may do as you will with me." The spirit said haughtily: "That is a very trifling feat," drew himself together, and made himself as small and slender as he had been at first, so that he crept through the same opening, and right through the neck of the bottle in again. Scarcely was he within than the boy thrust the cork he had drawn back into the bottle, and threw it among the roots of the oak into its old place, and the spirit was deceived.

AND NOW the schoolboy was about to return to his father, but the spirit cried very piteously: "Ah, do let me out! ah, do let me out!" "No," answered the boy, "not a second time! He who has once tried to take my life shall not be set free by me, now that I have caught him again." "If you will set me free," said the spirit, "I will give you so much that you will have plenty all the days of your life." "No," answered the boy, "you would cheat me as you did the first time." "You are spurning your own good luck," said the spirit; "I will do you no harm, but will reward you richly." The boy thought: "I will venture it, perhaps he will keep his word, and anyhow he shall not get the better of me." Then he took out the cork, and the spirit rose up from the bottle as he had done before, stretched himself out and became as big as a giant. "Now you shall have your reward," said he, and handed the boy a little rag just like sticking-plaster, and said: "If you spread one end of this over a wound it will heal, and if you rub steel or iron with the other end it will be changed into silver." "I must just try that," said the boy, and went to a tree, tore off the bark with his axe, and rubbed it with one end of the plaster. It immediately closed together and was healed. "Now, it is all right," he said to the spirit, "and we can part." The spirit thanked him for his release, and the boy thanked the spirit for his present, and went back to his father.

"Where have you been racing about?" said the father; "why have you forgotten your work? I always said that you would never come to anything." "Be easy, father, I will make it up." "Make it up indeed," said the father angrily, "that's no use." "Take care, father, I will soon hew that tree there, so that it will split." Then he took his plaster, rubbed the axe with it, and dealt a mighty blow, but as the iron had changed into silver, the edge bent: "Hi, father, just look what a bad axe you've given me, it has become quite crooked." The father was shocked and said: "Ah, what have you done? now I shall have to pay for that, and have not the wherewithal, and that is all the good I have got by your work." "Don't get angry," said the son, "I will soon pay for the axe." "Oh, you blockhead," cried the father, "wherewith will you pay for it? You have nothing but what I give you. These are students' tricks that are sticking in your head, you have no idea of woodcutting." After a while the boy said: "Father, I can really work no more, we had better take a holiday." "Eh, what!" answered he. "Do you think I will sit with my hands lying in my lap like you? I must go on working, but you may take yourself off home." "Father, I am here in this wood for the first time, I don't know my way alone. Do go with me." As his anger had now abated, the father at last let himself be persuaded and went home with him. Then he said to the son: "Go and sell your damaged axe, and see what you can get for it, and I must earn the difference, in order to pay the neighbor." The son took the axe, and carried it into town to a goldsmith, who tested it, laid it in the scales, and said: "It is worth four hundred talers, I have not so much as that by me." The son said: "Give me what you have, I will lend you the rest." The goldsmith gave him three hundred talers, and remained a hundred in his debt. The son thereupon went home and said: "Father, I have got the money, go and ask the neighbor what he wants for the axe." "I know that already," answered the old man, "one taler, six groschen." "Then, give him two talers, twelve groschen, that is double and enough; see, I have money in plenty," and he gave the father a hundred talers, and said: "You shall never know want, live as comfortably as you like." "Good heavens!" said the father, "how have you come by these riches?" The boy then told how all had come to pass, and how he, trusting in his luck, had made such a packet. But with the money that was left, he went back to the High School and went on learning more, and as he could heal all wounds with his plaster, he became the most famous doctor in the whole world. ●

Buddhist Economics

E.F. Schumacher

"RIGHT LIVELIHOOD" is one of the requirements of the Buddha's Noble Eightfold Path. It is clear, therefore, that there must be such a thing as Buddhist economics.

Buddhist countries have often stated that they wish to remain faithful to their heritage. So Burma: "The New Burma sees no conflict between religious values and economic progress. Spiritual health and material well-being are not enemies: they are natural allies."[1] Or: "We can blend successfully the religious and spiritual values of our heritage with the benefits of modern technology."[2] Or: "We Burmans have a sacred duty to conform both our dreams and our acts to our faith. This we shall ever do."[3]

All the same, such countries invariably assume that they can model their economic development plans in accordance with modern economics, and they call upon modern economists from so-called advanced countries to advise them, to formulate the policies to be pursued, and to construct the grand design for development, the Five-Year Plan or whatever it may be called. No one seems to think that a Buddhist way of life would call for Buddhist economics, just as the modern materialist way of life has brought forth modern economics.

Economists themselves, like most specialists, normally suffer from a kind of metaphysical blindness, assuming that theirs is a science of absolute and invariable truths, without any presuppositions. Some go as far as to claim that economic laws are as free from "metaphysics" or "values" as the law of gravitation. We need not, however, get involved in arguments of methodology. Instead, let us take some fundamentals and see what they look like when viewed by a modern economist and a Buddhist economist.

THERE IS universal agreement that a fundamental source of wealth is human labor. Now, the modern economist has been brought up to consider "labor" or work as little more than a necessary evil. From the point of view of the employer, it is in any case simply an item of cost, to be reduced to a minimum if it cannot be eliminated altogether, say, by automation. From the point of view of the workman, it is a "disutility"; to work is to make a sacrifice of one's leisure and comfort, and wages are a kind of compensation for the sacrifice. Hence the ideal from the point of view of the employer is to have output without employees, and the ideal from the point of view of the employee is to have income without employment.

The consequences of these attitudes both in theory and in practice are, of course, extremely far-reaching. If the ideal with regard to work is to get rid of it, every method that "reduces the work load" is a good thing.

The most potent method, short of automation, is the so-called "division of labor" and the classical example is the pin factory eulogized in Adam Smith's *Wealth of Nations*. Here it is not a matter of ordinary specialization, which mankind has practiced from time immemorial, but of dividing up every complete process of production into minute parts, so that the final product can be produced at great speed without anyone having had to contribute more than a totally insignificant and, in most cases, unskilled movement of his limbs.

The Buddhist point of view takes the function of work to be at least threefold: to give man a chance to utilize and develop his faculties; to enable him to overcome his egocenteredness by joining with other people in a common task; and to bring forth the goods and services needed for a becoming existence. Again, the consequences that flow from this view are endless. To organize work in such a manner that it becomes meaningless, boring, stultifying, or nerve-racking for the worker would be little short of criminal; it would indicate a greater concern with goods than with people, an evil lack of compassion and a soul-destroying degree of attachment to the most primitive side of this worldly existence. Equally, to strive for leisure as an alternative to work would be considered a complete misunderstanding of one of the basic truths of human existence, namely that work and leisure are complementary parts of the same living process and cannot be separated without destroying the joy of work and the bliss of leisure.

From the Buddhist point of view, there are therefore two types of mechanization which must be clearly distinguished: one that enhances a man's skill and power and one that turns the work of man over to a mechanical slave, leaving man in a position of having to serve the slave. How to tell the one from the other? "The craftsman himself," says Ananda Coomaraswamy, a man equally competent to talk about the modern West as the ancient East, "can always, if allowed to, draw the delicate distinction between the machine and the tool. The carpet loom is a tool, a contrivance for holding warp threads at a stretch for the pile to be woven round them by the craftmen's fingers; but the power loom is a machine, and its significance as a destroyer of culture lies in the fact that it does the essentially human part of the work."[4] It is clear, therefore, that Buddhist economics must be very different from the economics of modern materialism, since the Buddhist sees the essence of civilization not in a multiplication of wants but in the purification of human character. Character, at the same time, is formed primarily by a man's work. And work, properly conducted in conditions of human dignity and freedom, blesses those who do it and equally their products. The Indian philosopher and economist J. C. Kumarappa sums the matter up as follows:

If the nature of the work is properly appreciated and applied, it will stand in the same relation to the higher faculties as food is to the physical body. It nourishes and enlivens the higher man and urges him to produce the best he is capable of. It directs his free will along the proper course and disciplines the animal in him into progressive channels. It furnishes an excellent background for man to display his scale of values and develop his personality.[5]

If a man has no chance of obtaining work he is in a desperate position, not simply because he lacks an income but because he lacks this nourishing and enlivening factor of disciplined work which nothing can replace. A modern economist may engage in highly sophisticated calculations on whether full employment "pays" or whether it might be more "economic" to run an economy at less than full employment so as to ensure a greater mobility of labor, a better stability of wages, and so forth. His fundamental criterion of success is simply the total quantity of goods produced during a given period of time. "If the marginal urgency of goods is low," says Professor Galbraith in *The Affluent Society*, "then so is the urgency of employing the last man or the last million men in the labor force."[6] And again:

If . . . we can afford some unemployment in the interest of stability—a proposition, incidentally, of impeccably conservative antecedents—then we can afford to give those who are unemployed the goods that enable them to sustain their accustomed standard of living.

F ROM A BUDDHIST point of view, this is standing the truth on its head by considering goods as more important than people and consumption as more important than creative activity. It means shifting the emphasis from the worker to the product of work, that is, from the human to the subhuman, a surrender to the forces of evil. The very start of Buddhist economic planning would be a planning for full employment, and the primary purpose of this would in fact be employment for everyone who needs an "outside" job: it would not be the maximization of employment nor the maximization of production. Women, on the whole, do not need an "outside" job, and the large-scale employment of women in offices or factories would be considered a sign of serious economic failure. In particular, to let mothers of young children work in factories while the children run wild would be as uneconomic in the eyes of a Buddhist economist as the employment of a skilled worker as a soldier in the eyes of a modern economist.

While the materialist is mainly interested in goods, the Buddhist is mainly interested in liberation. But Buddhism is "The Middle Way" and therefore in no way antagonistic to physical well-being. It is not wealth that stands in the way of liberation but the attachment to wealth; not the enjoyment of pleasurable things but the craving for them. The keynote of Buddhist economics, therefore, is simplicity and non-violence. From an economist's point of view, the marvel of the Buddhist way of life is the utter rationality of its pattern—amazingly small means leading to extraordinarily satisfactory results.

For the modern economist this is very difficult to understand. He is used to measuring the "standard of living" by the amount of annual consumption, assuming all the time that a man who consumes more is "better off" than a man who consumes less. A Buddhist economist would consider this approach excessively irrational: since consumption is merely a means to human well-being, the aim should be to obtain the maximum of well-being with the minimum of consumption. Thus, if the purpose of clothing is a certain amount of temperature comfort and an attractive appearance, the task is to attain this purpose with the smallest possible effort, that is, with the smallest annual destruction of cloth and with the help of designs that involve the smallest possible input of toil. The less toil there is, the more time and strength is left for artistic creativity. It would be highly uneconomic, for instance, to go in for complicated tailoring, like the modern West, when a much more beautiful effect can be achieved by the skillful draping of uncut material. It would be the height of folly to make material so that it should wear out quickly and the height of barbarity to make anything ugly, shabby, or mean. What has just been said about clothing applies equally to all other human requirements. The ownership and the consumption of goods is a means to an end, and Buddhist economics is the systematic study of how to attain given ends with the minimum means.

MODERN ECONOMICS, on the other hand, considers consumption to be the sole end and purpose of all economic activity, taking the factors of production—land, labor, and capital—as the means. The former, in short, tries to maximize human satisfactions by the optimal pattern of consumption, while the latter tries to maximize consumption by the optimal pattern of productive effort. It is easy to see that the effort needed to sustain a way of life which seeks to attain the optimal pattern of consumption is likely to be much smaller than the effort needed to sustain a drive for maximum consumption. We need not be surprised, therefore, that the pressure and strain of living is very much less in, say, Burma than it is in the United States, in spite of the fact that the amount of labor-saving machinery used in the former country is only a minute fraction of the amount used in the latter.

Simplicity and non-violence are obviously closely related. The optimal pattern of consumption, producing a high degree of human satisfaction by means of a relatively low rate of consumption, allows people to live without great pressure and strain and to fulfill the primary injunction of Buddhist teaching: "Cease to do evil; try to do good." As physical resources are everywhere limited, people satisfying their needs by means of a modest use of resources are obviously less likely to be at each other's throats than people depending upon a high rate of use. Equally, people who live in highly self-sufficient local communities are less likely to get involved in large-scale violence than people whose existence depends on world-wide systems of trade.

From the point of view of Buddhist economics, therefore, production from local resources for local needs is the most rational way of economic life, while dependence on imports from afar and the consequent need to produce for export to unknown and distant peoples is highly uneconomic and justifiable only in exceptional cases and on a small scale. Just as the modern economist would admit that a high rate of consumption of transport services between a man's home and his place of work signifies a misfortune and not a high standard of life, so the Buddhist economist would hold that to satisfy human wants from faraway sources rather than from sources nearby signifies failure rather than success. The former tends to take statistics showing an increase in the number of ton/miles per head of the population carried by a country's transport system as proof of economic progress, while to the latter—the Buddhist economist—the same statistics would indicate a highly undesirable deterioration in the *pattern* of consumption.

ANOTHER striking difference between modern economics and Buddhist economics arises over the use of natural resources. Bertrand de Jouvenel, the eminent French political philosopher, has characterized "Western man" in words which may be taken as a fair description of the modern economist:

He tends to count nothing as an expenditure, other than human effort; he does not seem to mind how much mineral matter he wastes and, far worse, how much living matter he destroys. He does not seem to realize at all that human life is a dependent part of an ecosystem of many different forms of life. As the world is ruled from towns where men are cut off from any form of life other than human, the feeling of belonging to an ecosystem is not revived. This results in a harsh and improvident treatment of things upon which we ultimately depend, such as water and trees.[7]

The teaching of the Buddha, on the other hand, enjoins a reverent and non-violent attitude not only to all sentient beings but also, with great emphasis, to trees. Every follower of the Buddha ought to plant a tree every few years and look after it until it is safely established, and the Buddhist economist can demonstrate without difficulty that the universal observation of this rule would result in a high rate of genuine economic development independent of any foreign aid. Much of the economic decay of Southeast Asia (as of many other parts of the world) is undoubtedly due to a heedless and shameful neglect of trees.

MODERN ECONOMICS does not distinguish between renewable and non-renewable materials, as its very method is to equalize and quantify everything by means of a money price. Thus, taking various alternative fuels, like coal, oil, wood, or water-power: the only difference between them recognized by modern economics is relative cost per equivalent unit. The cheapest is automatically the one to be preferred, as to do otherwise would be irrational and "uneconomic." From a Buddhist point of view, of course, this will not do; the essential difference between non-renewable fuels like coal and oil on the one hand and renewable fuels like wood and water-power on the other cannot be simply overlooked. Non-renewable goods must be used only if they are indispensable, and then only with the greatest care and the most meticulous concern for conservation. To use them heedlessly or extravagantly is an act of violence, and while complete non-violence may not be attainable on this earth, there is nonetheless an ineluctable duty on man to aim at the ideal of non-violence in all he does.

Just as a modern European economist would not consider it a great economic achievement if all European art treasures were sold to America at attractive prices, so the Buddhist economist would insist that a population basing its economic life on non-renewable fuels is living parasitically, on capital instead of income. Such a way of life could have no permanence and could therefore be justified only as a purely temporary expedient. As the world's resources of non-renewable fuels — coal, oil and natural gas — are exceedingly unevenly distributed over the globe and undoubtedly limited in quantity, it is clear that their exploitation at an ever-increasing rate is an act of violence against nature which must almost inevitably lead to violence between men.

This fact alone might give food for thought even to those people in Buddhist countries who care nothing for the religious and spiritual values of their heritage and ardently desire to embrace the materialism of modern economics at the fastest possible speed. Before they dismiss Buddhist economics as nothing better than a nostalgic dream, they might wish to consider whether the path of economic development outlined by modern economics is likely to lead them to places where they really want to be. Towards the end of his courageous book *The Challenge of Man's Future*, Professor Harrison Brown of the California Institute of Technology gives the following appraisal:

Thus we see that, just as industrial society is fundamentally unstable and subject to reversion to agrarian existence, so within it the conditions which offer individual freedom are unstable in their ability to avoid the conditions which impose rigid organization and totalitarian control. Indeed,

when we examine all of the foreseeable difficulties which threaten the survival of industrial civilization, it is difficult to see how the achievement of stability and the maintenance of individual liberty can be made compatible.[8]

Even if this were dismissed as a long-term view there is the immediate question of whether "modernization," as currently practiced without regard to religious and spiritual values, is actually producing agreeable results. As far as the masses are concerned, the results appear to be disastrous—a collapse of the rural economy, a rising tide of unemployment in town and country, and the growth of a city proletariat without nourishment for either body or soul.

It is in the light of both immediate experience and long-term prospects that the study of Buddhist economics could be recommended even to those who believe that economic growth is more important than any spiritual or religious values. For it is not a question of choosing between "modern growth" and "traditional stagnation." It is a question of finding the right path of development, the Middle Way between materialist heedlessness and traditionalist immobility, in short, of finding "Right Livelihood." ●

NOTES

1. *The New Burma* (Economic and Social Board, Government of the Union of Burma, 1954).
2. *Ibid.*
3. *Ibid.*
4. *Art and Swadeshi* by Ananda K. Coomaraswamy (Ganesh & Co., Madras).
5. *Economy of Permanence* by J.C. Kumarappa (Sarva-Seva Sangh Publication, Rajghat, Kashi, 4th edn., 1958).
6. *The Affluent Society* by John Kenneth Galbraith (Penguin Books Ltd., 1962).
7. *A Philosophy of Indian Economic Development* by Richard B. Gregg (Navajivan Publishing House, Ahmedabad, India, 1958).
8. *The Challenge of Man's Future* by Harrison Brown (The Viking Press, New York, 1954).

Threepenny Bit

P.L. Travers

IN THE DAYS when I was not much taller than a well-grown daisy, one did not talk about money. It was not a suitable subject for conversation. If you lost a milk tooth you did not, like modern children, put it under your pillow so that the fairies could turn it into copper or silver. The fairies we knew would never have thought of such a thing—the vulgarity of it!

True, my Irish father carried on his watch-chain a small square contraption which, if you pressed one button, something called a sovereign rose up out of its metal bed and, if another, a half-sovereign leapt from its place, both of them made of gold. This, I thought, was a kind of toy, as it might be a jack-in-the-box. I had no idea that it was money, nor indeed what money was. Living as we did in a small country settlement in the semi-tropical Antipodes, there was nothing to which the word related except a small corrugated tin building where a conglomeration of articles could be exchanged—for what? I was soon to know. There were packets of daily edibles, salt, sugar, flour, tea, but to us those things came in bushel bags; there were reels of cotton, pins and needles, but to us these referred to the dressmaker who periodically spent some weeks in the house and departed, leaving behind her new garments of every kind.

Only when they discovered among the crowding merchandise a shelf of slender paper books—green for a fairy tale, blue-and-red for a Buffalo Bill—did my parents give me every week a round copper object known as a penny for which I could exchange or buy—a new word!—one of those delectable objects, so much more interesting than the books on my father's shelves.

And, of course, there were all those little discs of silver and occasionally a slip of colored paper that were known collectively, I learned, as money, that appeared in the collection plate on Sundays in the little rustic church. What did God do with them, I wondered. He owned the world—was not that enough? I tried once to hold on to the contribution that was regularly slipped into my palm, but my Atlas-eyed father, lustily singing an altered version of a hymn I knew well:

Time like an ever-rolling stream
Put in that sixpence now!

shocked the coin out of my hand. God
got it, with the rest.

BUT THE TIME CAME when I
had some money of my
own. While waiting for a governess
hardy enough to endure living so far
behind the back of beyond, I was sent
to the local school. Oh, the lovely
multiplication tables, each chanted to a
different note of the octave, and the
children riding their ponies in from
the outback!

There was only one teacher and
it so happened that she lost her brooch
and I found it. For this I was given a
big round silver coin that she called a
two-shilling-piece. I took it gladly
and told no one. Hadn't I learned that
money was not a subject for social
conversation?

But somehow the news got out
and my father gave vent to his favorite
impersonation of Zeus in one of his
rages.

"That a child of mine"—my
mother, apparently, had no part in
me—"should accept money—money!—
for performing a courteous act! Where
is it?"

"I spent it at the tin shop."
"On what?"
"Cigarettes."
! ! !
For a moment, he had no words.
"You smoked them?" Really, if
at seven-and-a-half I could do such a
thing, where would I be at seventy?
"No, I ate them."
"Ye gods, a tobacco chewer!"
"No, no!" my intuitive Scottish
mother made a dart at my pinafore
pocket and triumphantly brought forth
a dilapidated package. It was labeled
Simpson's Sugar Smokes and in it there
was one lone stick of marzipan, the
white reddened with cochineal at one
end.

Hadn't I learned that money was not a subject for social conversation?

I had met money. I knew what it
was to go shopping. There was nothing they could do about it, nothing
even my father could say and very little time to say it, if he could have
found the words. At the age of forty-
three, riding all day in a tropical
downpour, he came home drenched to
the skin and contracted what was
called galloping pneumonia. The
house was silent, drawn into itself,
and outside, for he was universally
loved, horses were everywhere as people rode in for news of him.

I was taken to see him, white as
his pillows, and opened my fist to
show him what remained of my
spending spree—a threepenny bit.

"I am going to buy you some
pears," I told him and waited, almost
hoping, for Zeus to descend. But he
took it gently and gratefully and
slipped it under his pillow. "Pears are
exactly what I need," he said, as
though I had brought him the elixir of
life. But it was another elixir, for the
next day, my mother, looking childish
and forlorn, told me that Father had
gone to God. I did not believe her. He
was not a man to do a thing like that.
And, anyway, God had all those an-

*In its essence it
has become part of
myself, often at odds
with the other parts.*

---

gels. He did not need my father. It took me years to accept it.

A S FOR the threepenny bit, it remained for a time, nestling in one of the velvet crevices in my mother's jewel case, infinitely precious as it was the last thing he touched. Where is it now, after all these years? Probably spent by someone not knowing its history and before it went out of currency. But in its essence it has become part of myself, often at odds with the other parts, and an instrument of fate. It carried us—a young mother with three small children—far away south to temperate zones where we, too, lived a life in which sugar came in little packets and bread from a baker's basket instead of our own oven. Money was discussed freely—no longer because of its existence but because, after my father's departure and at the same time a crash (whatever that was) that took my mother's fortune—it was in short supply.

It is those who have to be careful with it who need to talk about money. All this, however, did not upset the children. Since they had never had it

in esse, merely what it brought, they unconditionally accepted the new countryside and the little town a mile away. There was a new world to be discovered. Only to me was it gradually made clear by the rich relatives who sat like crows on the fence of our horizon that, being the eldest, my purpose in life was to help my mother, not merely by washing a dish or making a bed but by working—by making money—a quiet, insignificant, undemanding life.

Little did they guess that by sending me to an elegant boarding school—the only one they knew—where I could read and learn, they were preparing me for my own purpose that, ever since I could remember, had been to transport myself to the place my parents had always spoken of—so, indeed, did all their generation—as Home, which meant crossing the world. It was there that I would find the answers to the questions that were in me.

B UT RICH RELATIVES are only minimally kind to poor relations. The crows on the fence, sitting on their golden eggs, hearing that I had passed the university entrance a year younger than was necessary, were happy to save fees and did what they had always intended. They had me taught typing and used their influence with the head of a large prestigious firm in a southern city to persuade him to offer me a job.

He did not relish the idea. "No, no, she is far too young"—I was in my school uniform—"and it would not be appropriate for this kind of girl."

"Nonsense, Sir Thomas," said my great-aunt. "She's perfectly competent and she needs to work." And since few people ever dared to outface that lady, Sir Thomas had to give in.

Luckily I was not placed under

the aegis of the Gorgon who managed the general bevy of typists, but dumped into the cashier's office where I was the only one. My typewriter was the kind that adds up figures and probably because of my early association with money, or rather the lack of it, I never once got them right. Some man, patiently accepting the inevitable, did the whole thing over every day on an old-fashioned adding machine. No word of this was bruited to the management. I was even recommended for a raise, which made my great-aunt preen herself on being right again.

Thus I spent my days, and at night sat up in my bed writing, till it happened that a friend saw my notebook and took it to an editor who offered to print the contents and, to my shocked surprise, pay me money for them. It seemed to me wrong, at this first contact, that something which I thought of as a gift should be exchanged for silver and gold. I had to learn, and the threepenny bit backed me up, reminding me that if I wanted to fulfill my purpose I had to build up funds. And at last it happened! I left the typewriter that never gave the right answer and, to the dismay of my family, joined the other world.

"So uncertain!" said my mother, sadly.

"Preposterous! Most unsuitable!" said my great-aunt.

And later, when I had met many who urged me on to my purpose, "And how, pray, will your mother manage, if you go to England?"

But I had it all worked out. I would leave behind me enough money to cover the six weeks' voyage and a further six weeks during which time I was certain I would get a job.

So, though uneasily, I was allowed to depart on condition that I was met by and stayed with specific members of the family. This happened as they had hoped. I was welcomed to a big house—with three cars—just outside London. The relatives were just going off on a trip to Cannes. I would go with them, they decided, returning in good time for the Season, become the daughter they had never had, and give up "all this foolish talk of art and literature."

I cried off Cannes, saying I would like to get the feel of England and while they were away presented various letters of introduction editors had given me. By the time they returned I had found a job and a place

to live, though not the attic I had romantically expected and the threepenny bit had determined on.

All this led eventually to my ultimate goal, which was Ireland. A poem did it, bringing me to A.E. and Yeats, Dublin with a poet on every corner, and so to my father's family home, away in the long blue hills. Lovers of horses and the countryside, they were not so in love with Caitlin ni Houlihan as he had been and they did not at all approve of me "going about in Dublin with men who see fairies."

The question of money never arose—it seldom does in Ireland—even if, for all I knew, it was needed. But the threepenny bit found a sibling there. Whatever my uncle's private charities may have been, it was household rule that nobody spent a sixpenny piece, not even the servants, who in Ireland are accepted as part of the family. If such a thing appeared as change in any transaction it was put in a bag that hung on the inside of the hall door and on Saturdays, needy local women could be seen traipsing up the drive to have the contents shared among them.

So it was that I found the psychological roots that unknowingly I had always had. From that safe base, not neglecting my responsibilities, I could go wherever I wished. And as time went on I travelled much of the world. I was even asked by an anthropologist to visit the Eskimos and gladly would have gone to listen to their stories had he not warned me that I should be living largely on sealblubber. No amount of dollars would buy me what, anyway, was not to be had, a piece of bread and butter!

THERE IT WAS again— money! This commodity which seems to beggar those who possess it (even millionaires never have enough) and those who don't. If only we could return to cowrie shells! For me, the tussle between my usual self and the threepenny bit has continued all my grown-up life, a silent shouting match between two of the same gender. I could give even a small amount, and everything was peaceful, more so the more I gave. But to spend on myself always put me in question. Was it necessary? Why couldn't I manage without it? Remember, "Enough is too much!"

Once—oh, it was long ago—I designed and had made a fur coat. If it had been rabbit my threepenny bit would probably not have minded; rabbit was thought of as vermin in my childhood, even though people ate it. But it was Persian lamb, and in it I strutted around Fifth Avenue closing my ears to the threepenny bit's shrieks of rage.

Recently, no longer able to bear the shame that the memory now evokes, I offered it to Oxfam. They refused. "We no longer take fur coats," they said, and I loved them for it in spite of my self-abasement. They would not feed the world's newborn children from money—and be sure there would be someone to offer it— paid for the skins of little unborn lambs.

HOW CAN WE "repair the past" and so "prepare the future?" In a time of growing materiality and cultural decline, this question has arisen as a reconciling factor between myself and the threepenny bit. As the sky darkens and a summer day in November makes one think of the ozone layer rather than rejoicing; when the rich get richer and the poor sleep in doorways; when cyclists wear masks as they ride through the traffic; when pamphlets made from precious forest trees pour in a stream through the letter box offering prizes, trips abroad, new sources for borrowing money, and, as well, cries for help from everywhere, I sometimes feel like throwing all these papers on to the pile of things to be recycled.

"But," I protest to the threepenny bit, "I have already saved the elephants, the dolphins, the whales and the seals." She, however, is adamant, making me feel guilty, not only for the plight of the creatures but for the demise of mammoth, brontosaur, and dodo.

"And, if you throw the requests away, what about the children?" she asks, slyly, knowing well I could never say No to them whatever the size of my pittance. If only it were a procedure on a human scale—of buying pears, say, for someone one loves. Or, on the other hand, a cosmic matter—earthquake, wind or rainbow, the things we cannot help! But is that true? How can we know what effect we have upon the Earth? Did not man invent the atom bomb? So, I cannot escape the conviction that the springs of our present plight are in myself. As without, so within. In the microcosm of our own beings one finds the poison, the things that are out of joint. It is we who have so wounded Gaia—may it not be mortally.

In quiet moments of togetherness, and such moments come more frequently now, I find myself quoting Plato to the threepenny bit. Today I recited the prayer of Socrates that comes at the end of the *Phaedrus*.

"Beloved Pan, and all ye other Gods that here abide, grant me to be peaceful in the inner man and all that I have of outer things to be at peace with those within. May I count only the wise man rich and my store of gold be such as none but the good can bear."

And her comment, gentle and wise, was "Yes,"—a rare word for her—"We must redeem the time." ●

SANIE HOLLAND

ARCS

The Uses of Money

MY DADDY is dead, but I don't know how;
He left me six horses to follow the plow.

I sold my horses to buy me a cow,
Now wasn't that a pretty thing to follow the plow?

I sold my cow for to buy me a calf;
I never made a bargain but I lost the better half.

I sold my calf for to buy me a cat,
To sit before the fire and to warm her pretty back.

I sold my cat for to buy me a mouse;
It set fire to its tail and burned down the house.

—Children's Song

WAR BEGETTETH poverty; poverty, peace.
Peace maketh riches flourish; wonders never cease.
Riches beget pride; pride is war's best ground.
War begetteth poverty; the world goeth round.

—English Round

I GAVE MY THORN and got some bread.
I lost the bread and gained a hen.
I gave the hen for some nice red yarn.
I traded the yarn for a fine old coat.
I gave the coat to get a dog
and swapped the dog for a nice fat lamb.
I traded the lamb for a mandolin,
and this I will keep to play my songs.[1]

— The Cat Who Loved to Sing

W ITH MONEY in your pocket, you are wise and you are handsome and you sing well too.[2]

— *Yiddish Proverb*

A T LOW WATER, I went on board, and though I thought I had rummaged the cabin so effectually as that nothing more could be found, yet I discovered a locker with drawers in it, in one of which I found two or three razors, and one pair of large scissors, with some ten or a dozen of good knives and forks; in another, I found about thirty-six pounds value in money, some European coin, some Brazil, some pieces of eight, some gold, some silver.

I smiled to myself at the sight of this money. "O drug!" said I aloud, "what art thou good for? Thou art not worth to me, no, not the taking off of the ground; one of those knives is worth all this heap. I have no manner of use for thee; even remain where thou art, and go to the bottom as a creature whose life is not worth saving."[3]

— *The shipwrecked Robinson Crusoe*

To the question, [the Psalmist] makes reply: *The light of thy counte-nance, O Lord, is signed upon us.* This light which shines upon the mind, not upon the eyes, constitutes mankind's whole and essential good. It is *signed upon us,* so the Psalmist has said, like a coin stamped with the king's image. For man was made to God's image and likeness, and defaced it by sin. His true and lasting good therefore is to be stamped anew by regeneration. This seems to me the sense which wise interpreters have applied to our Lord's words upon looking at Caesar's tribute money: *Render to Caesar the things that are Caesar's, and to God the things that are God's.* It is as if He had said: God, like Caesar, demands from you the impression of His own image. Just as you repay his coinage to Caesar, so return your soul to God, shining and stamped with the light of His countenance.[4]

—St. Augustine

The wise and moral man
Shines like a fire on a hilltop,
Making money like the bee,
Who does not hurt the flower.[5]

— The Pali Canon

Money is like a sixth sense without which you cannot make a complete use of the other five.[6]

—William Somerset Maugham

Put not your trust in money, but put your money in trust.[7]

—Oliver Wendell Holmes

If a man is wise, he gets rich, an' if he gets rich, he gets foolish, or his wife does. That's what keeps money movin' around.[8]

—Finley Peter Dunne

There are people who have money and people who are rich.

—Coco Chanel

One can relish the varied idiocy of human action during a panic to the full, for, while it is a time of great tragedy, nothing is being lost but money.[9]

—John Kenneth Galbraith

ANOTHER QUESTION preoccupied us a great deal during those days of waiting. You do not set out to acquire something in a foreign country without a supply of money. For purposes of barter with such "savages" and "natives" as they meet, explorers usually carry with them all sorts of knickknacks and junk—knives, mirrors, Paris souvenirs, inventors' leftovers, braces, patent garters, trinkets, cretonne, pieces of soap, eau-de-vie, old rifles, harmless ammunition, saccharin, caps, combs, tobacco, pipes, medals, and cord—not to speak of devotional objects. Since we might, during the course of the voyage and even in the interior of the continent, meet people belonging to ordinary humanity, we had provided ourselves with such wares. But in our relations with the superior beings of Mount Analogue, what would be suitable for barter? What did we possess of real value? With what could we pay for the new knowledge we were seeking? Would we have to accept it as charity or on credit?

Each one of us kept making his personal inventory, and each one of us felt poorer as the days went on. For no one saw anything around him or in him which really belonged to him. It reached the point where we were just eight beggars, possessing nothing, who each night watched the sun sink toward the horizon.[10]

—René Daumal

NOTES

1. From Nonny Hogrogian, *The Cat Who Loved to Sing* (New York: Alfred A. Knopf, 1988).
2. From *Yiddish Proverbs*, edited by Hanan J. Ayalti (New York: Schocken Books, 1949).
3. From Daniel Defoe, *The Life and Strange Surprising Adventures of Robinson Crusoe, of York, Mariner, as Related by Himself* (London: Henry Frowde and Hodder & Stoughton, 1910).
4. From *St. Augustine on the Psalms*, Dame Scholastica Hebgin and Dame Felicitas Corrigan, trans. (New York: Newman Press, n.d.).
5. From *The Pali Canon*, Singalavada-Sutta, Digha-nikaya, 3:180, translated by William Theodore de Bary (o.p.).
6. From William Somerset Maugham, *Of Human Bondage* (Salem, N.H.: Ayers Co. Publishers, 1977), Ch. 51.
7. From Oliver Wendell Holmes, *The Autocrat of the Breakfast Table* (Arden Library, 1977).
8. From Finley Peter Dunne, *Observations by Mr. Dooley* (Westport, Conn.: Greenwood Press Inc., n.d.), "Newport."
9. From John Kenneth Galbraith, *The Great Crash, 1929* (New York: Avon Books, 1980), ch. 1.
10. From René Daumal, *Mount Analogue* (New York: Penguin, 1974).

The Degeneration of Coinage

René Guénon

I F THE MERELY "economic" point of view as it is understood today is not departed from, it certainly seems that money is something which appertains as completely as possible to the "reign of quantity." This indeed is the reason why it plays so predominant a part in modern society, as is only too obvious, and it would clearly be superfluous to insist on that point; but the truth is that the "economic" point of view itself, and the exclusively quantitative conception of money which is inherent in it, are but the products of a degeneration which is on the whole fairly recent, and that money possessed at its origin, and retained for a long time, quite a different character and a truly qualitative value, remarkable as this may appear to the majority of our contemporaries.

It may very easily be observed, provided only that one has "eyes to see," that the ancient coins are literally covered with traditional symbols, often chosen from among those which carry some particularly profound meaning; thus for instance it has been observed that among the Celts the symbols figured on the coins can only be explained if they are related to the doctrinal knowledge which belonged to the Druids alone, and this implies a direct intervention of the Druids in the monetary domain. There is not the least doubt that the truth in this mat-

ter is the same for the other peoples of antiquity as for the Celts, of course after taking account of the modalities peculiar to their respective traditional organizations. This is fully in agreement with the fact of the inexistence of the profane point of view in strictly traditional civilizations: money itself, where it existed at all, could not be the profane thing it came to be later; and if it had been so, how could the intervention of a spiritual authority, which would then obviously have no concern with money, be explained, and how would it be possible to understand that many traditions speak of coinage as of something really charged with a "spiritual influence," the action of which could not become effective except by means of the symbols which constituted its normal "support"? It may be added that right up to very recent times it was still possible to find a last vestige of this notion in devices of a religious character, which certainly retained no real symbolical value, but were at least something like a recollection of the traditional idea, more or less uncomprehended thenceforth; but after having been relegated in certain countries to a place round the rim of coins, in the end these devices disappeared completely, indeed there was no longer any reason for them as soon as the coinage represented nothing more than a "material" and quantitative token.

THE CONTROL of money by the spiritual authority, in whatever form it may have been exercised, is by no means exclusively confined to antiquity, for, without going outside the Western world, there is much to indicate that it must have been perpetuated until towards the end of the Middle Ages, that is, for as long as the Western world had a traditional civilization. It is impossible to explain in any other way the fact that certain sovereigns were accused at this time of having "debased the coinage"; since their contemporaries regarded this as a crime on their part, it must be concluded that the sovereigns had not the free disposal of the standard of the coinage, and that, in changing it on their own initiative, they overstepped the recognized rights of the temporal power.[1] If that were not the case, such an accusation would have been quite without meaning; the standard of the coinage would only then have had an importance based on convention, and it would not have mattered, broadly speaking, if it has been made of any sort of metal, or of various sorts, or even been replaced by mere paper as it is for the most part today, for this would have been no hindrance to the continuance of exactly the same "material" employment of it. An element of another order

must therefore have been involved, and it must have been of a superior order, for unless that had been the case the alteration could not have assumed a character so exceptionally serious as to end in compromising the very stability of the royal power; but the royal power by acting in this way, usurped the prerogatives of the spiritual authority, which is without any doubt the one authentic source of all legitimacy.

In this way the facts, which profane historians seem scarcely to understand, conspire once more to indicate very clearly that the question of money had in the Middle Ages as well as in antiquity aspects quite unknown to the moderns.

WHAT HAS HAPPENED in this case is but an example of a much more general happening, affecting all activities in every department of human existence; all have been gradually divested of any "sacred" or traditional character, and thereby that existence itself in its entirety has become completely profane and is now at last reduced to the third-rate mediocrity of "ordinary life" as it is found today. At the same time, the example of money clearly shows that this "profanization"—if any such neologism be allowable—comes about chiefly by the reduction of things to their quantitative aspect alone; indeed, nobody is able any longer to conceive that money can represent anything other than a simple quantity; but, although the case of money is particularly apt in this connection because it has been as it were carried to the extreme of exaggeration, it is very far from being the only case in which a reduction to the quantitative can be seen as contributing to the enclosure of existence within the limited horizon of the profane point of view. This is sufficiently understandable after what has been said of the

peculiarly quantitative character of modern industry: by continuously surrounding man with the products of that industry, and so to speak never letting him see anything else (except, as in museums for example, in the guise of mere "curiosities" having no relation with the "real" circumstances of his life and consequently no effective influence on it), he is really compelled to shut himself up inside the narrow circle of "ordinary life," as in a prison without escape. In a traditional civilization, on the contrary, each object was at the same time as perfectly fitted as possible for the use for which it was immediately destined and also made so that it could at any moment, and owing to the very fact that real use was being made of it (instead of its being treated more or less as a dead thing as the moderns do with everything that they consider to be a "work of art"), serve as a "support" for meditation, linking the individual with something other than the mere corporeal modality, thus helping everyone to elevate himself to a superior state according to the measure of

his capacities:[2] what an abyss there is between these two conceptions of human existence!

The qualitative degeneration of all things is closely linked to that of money, as is shown by the fact that nowadays the "worth" of an object is ordinarily "estimated" only in terms of its price, considered simply as a "figure," a "sum" or a numerical quantity of money; in fact, with most of our contemporaries, every judgment brought to bear on an object is nearly always based exclusively on what it costs. The word "estimate" has been emphasized because it has in itself a double meaning, qualitative and quantitative; today the first meaning has been lost to sight, or, what amounts to the same thing, means have been found to equate it to the second, and thus it comes about that not only is the "worth" of an object "estimated" according to its price, but the "worth" of a man is "estimated" according to his wealth.[3] The same thing has naturally happened to the word "value," and it may be noticed in passing that on this is based a curious abuse of the word by certain recent philosophers, who have even gone so far as to invent as a description of their theories the expression "philosophy of values"; underlying their thoughts is the idea that everything, to whatever order it may belong, is capable of being conceived quantitatively and expressed numerically; and "moralism," which is their other predominant preoccupation, thus comes to be closely associated with the quantitative point of view.[4] These examples show too that there has been a real degeneration of language, inevitably accompanying or following that of everything else; indeed, in a world in which every attempt is made to reduce all things to quantity it is evidently necessary to use a language which itself evokes nothing but purely quantitative ideas.

To RETURN more particularly to the question of money, one more point remains to be dealt with, for a phenomenon has appeared in this field which is well worthy of note, and it is this: since money lost all guarantee of a superior order, it has seen its own actual quantitative value, or what is called in the jargon of the economists its "purchasing power," becoming ceaselessly less and less, so that it can be imagined that, when it arrives at a limit which is getting ever nearer, it will have lost every justification for its existence, even all merely "practical" or "material" justification, and that it will disappear of itself, so to speak, from human existence. It will be agreed that here affairs turn back on themselves in a curious way, but the preceding explanations will make the idea quite easy to understand: for since pure quantity is by its nature beneath all existence, when the trend towards it is pressed to its extreme limit, as in the case of money (more striking than any other because the limit has nearly been reached), the end can only be a real dissolution. The case of money alone already shows clearly enough that, as was said above, the security of "ordinary life" is in reality a highly precarious thing, precarious in many other respects as well; but the positive conclusion that will emerge will be always the same, namely, that the real goal of the tendency which is dragging men and things towards pure quantity can only be the final dissolution of the present world.　　　　●

1. See *Autorité spirituelle et pouvoir temporel*, p. 111, where the case of Philippe le Bel is specially referred to, and where it was suggested that there may be a fairly close connection between the destruction of the Order of Templars and the alteration of the coinage, and this is easily understood if it is recognized as at least very plausible that the Order of Templars then had the function among others, of exercising spiritual control in this field; the matter need not be pursued further here, but it may be recalled that the beginning of the modern deviation properly so-called has been assigned precisely to this moment.
2. Numerous studies by A.K. Coomaraswamy may be consulted on this subject, which he has developed profusely and "illustrated" in all its aspects with all necessary explanations.
3. The Americans have gone so far in this direction that they commonly say that a man is "worth" so much, intending to convey in that way the figure to which his fortune has risen; they say too, not a man has succeeded in his affairs, but that he "is a success," and this is as much as to identify the individual completely with his material gains.
4. This association by the way is not an entirely new thing, for it actually goes back to the "moral arithmetic" of Bentham, which dates from the end of the eighteenth century.

Excerpted from René Guénon, *The Reign of Quantity* (London: Luzac & Company, Ltd., 1953).

EPICYCLE

The Ninety-Nine Rupee Snare/Hindu

THE MERCHANT'S WIFE sat at her husband's feet massaging his legs when she started in on him again. "Do you ever look in the mirror?" she began. "What a state you are in! Not even forty years old and you look more than sixty. You can't sleep at night. You look malnourished. You are worth millions. You have a flourishing business. Two grown sons. Two lovely daughters-in-law. Everything a person could want, but day and night, night and day, you go about with your face hanging to the ground. I cannot understand what troubles stir inside you. Always to have that morose expression on your face! What is it? Are you saving up money to take to your next life? You mumble about money even in your sleep. You should just burn it all if it causes so much grief. I look into your eyes and feel my insides wither. Look at the cotton carder who lives next door. Every day he spends bent over his cotton, beating and combing, and still he seems content. Strong, and sturdy as a horse. He's already over seventy years old, but looks thirty. Not a white hair in his head. Labors all day so that he can scrape together a meal at night. And still he never worries about tomorrow. He sleeps his own sleep, and wakes at his own hour. He and his wife are both happy. We splash around in millions and yet they seem better off than we."

ROBIN A. SMITH

The merchant was so engrossed in his accounts he heard only half her nonsense. But when she came to the bit about the cotton carder he was unable to contain himself. He said, "That old Ninety-Nine Rupee Snare hasn't gotten a hold of him yet. But the Snare's maw is treacherous. Once he gets caught, his idyllic happiness will fly with the breeze. Who knows better than I? You only need to ask and I will show you how bad that Snare can be. It may mean losing ninety-nine rupees, but then you will understand my bind and quit your ceaseless nagging. Fill a sack with ninety-nine rupees and throw it into his yard. Then wait a few days," he told his wife. "If this innocent happiness of his does not become a thing of the past, then I will eat my name."

"Not everyone in this world is like you," his wife interjected. "Finding the money will only heighten his present joy. He starts to work when the sun is high and leaves off once evening comes. This money will only make him more at ease. He never worries about his tomorrows anyway."

"Seems I will have to lose ninety-nine rupees."

"Lose? What loss? He will return the money as soon as he finds it. He is not the type to keep anything that is not rightly his. He is satisfied with the fruits of his own honest labor."

THE ARGUMENT between the merchant and his wife brought about an unusual turn of events. The next morning when the carder's wife was sweeping out their yard she came across a sack filled with coins.

She looked to the right, and looked to the left. She looked up, and she looked down. Then she marched straight over to her husband and shook him awake. "Look," she demanded, "This sack must certainly belong to the merchant. Should I go return it? In all these years we have not given them even one opportunity to complain. Now we will be blamed for something that we did not even do. Our reputation will be ruined. I tell you this: he is not the type to keep quiet. He will cause quite a scene."

The carder listened to her with astonishment. What a lucky day! A bag filled with coins was found in our own yard! How could this be called thievery? "Count it first," he told her. "How can we return it without counting it?"

They opened the sack together and started to count the coins. One, two, three, . . . until they reached ninety-nine.

"Ninety-nine?!" the carder said in surprise. "Perhaps we miscounted. There should be one hundred. Let's try counting again."

They counted all the coins again, then a third time, and a fourth. But there was never one hundred, nor could there ever be. They searched all through the yard, but it was useless. Another coin was not to be found.

The carder put the money back into the sack and suggested they wait until evening. "We didn't set out to steal it, after all," he decided. "Who knows, it could have been dropped by some bird or bug flying over. If this money does belong to the merchant he certainly will come over and claim it."

The carder's wife nodded her head. This seemed to suit her as well.

THAT DAY the merchant's wife called her neighbor over again and again on some pretext or other. First she needed a pillow stuffed, then a mattress, then a quilt. But the carder's wife never once talked about the money that was found. Was her husband right after all?

Three days passed, and still no mention was made of the sack of coins. In one house the merchant was taunting his wife about the return of the sack, while in the other house the cotton-carder was thinking of ways to add another rupee to the sack so that he would have a hundred.

Now for the first time the rhythmic "dhoon-dhat, dhoon-dhat" sound of carding carried on past the sinking of the sun and deep into the night. The merchant rolled over in bed and exclaimed, "This is no dream. Listen! It's midnight and the carder is still at work! It can't be!"

"Hush," his wife said, "And lie still. Why are you complaining? If someone wants to stay up late working, why should it bother you? Be quiet and go to sleep."

The next day hours before the rising of the sun the carder's melody could be heard again: dhoon-dhat, dhoon-dhat. Well past dawn and much before dusk he could be seen toiling, hour upon hour. Finally on the seventh day he was able to add another rupee to the sack. Now he had a hundred!

The husband and wife beamed with pleasure. They sat down and counted the money, over and over. How pleasing was the sound of jingling coins! A far sweeter melody than the dhoon-dhat of the card.

The carder began to fantasize ways of increasing his wealth. A hundred . . . thousand . . . ten thousand . . . then he would have a million in no time. He would card the cottony dreams of the universe. He would card the stuff of the sun, the moon, and the countless stars. With hard work and dedication, anything is possible.

Once wealthy he would buy an elephant and ride upon a saddle of gold. He would sit up there all day, carding his cotton and riding around the town. He would accept orders only from the king and queen. For stuffing a pillow, he would accept payment only in gold. It was theirs to decide. And if the cotton itself were to turn to spun gold . . .

EVEN IN DAYLIGHT his gnarled cotton began to take on a golden hue. And now the dhoon-dhat of his card could be heard in the middle of the night. He would go for five days at a time without a wink of sleep. And he was never in the mood to eat.

He starved day after day, and toiled night after night, and saved another hundred rupees in no time. Two hundred now! But it was not long before he began to look old and feeble. His hair turned white. His cheeks became gaunt and his eyes sunken. His back became crooked, and his lips chapped. Even so, he would not sit idle for a moment. He started cheating his customers as he cleaned and plumped the batting.

The claws of greed are sharper than those of death. All his calm and contentment had flown with the breeze. Even his precious teeth began dropping out one by one. Yet, he felt happy. The joy of amassing wealth was surely one of the greatest joys known to man! The cotton-carder no longer dreamed his dreams while fast asleep, but dreamed his dreams with eyes wide open. ●

From a 1982 collection of stories by Vijay Dan Detha, English translation © 1990 by Christi Ann Merrill.

Four Indian Attitudes Toward Money

Karl H. Potter

THE LOVE OF money is, according to the Christian New Testament, the root of all evil. That is a typically Western point of view, supported by the amount of attention given to acquiring material wealth, particularly in American society. It is interesting to compare this form of attachment with the "four aims of life" in traditional Hindu thought, and thus to relate a specific viewpoint toward money to the implications of a standard classification of human concerns handed down for over two millennia.

When reflecting on man's situation the classical thinkers of ancient India had recourse to a series of four notions frequently called, in contemporary accounts of them, "aims of life." The Sanskrit terms for these four notions are *artha, kāma, dharma,* and *mokṣa.* The proper understanding of the meaning of these terms, though essential for understanding classical Indian philosophies, is by no means easy.

To call these four things "aims"

It is the attitude of minimal concern which is the attitude marked out by the word artha.

suggests that they are *states* toward which one aims. Now in some sense perhaps the last of the four, *mokṣa* or complete freedom, is a state, but the sense in which this is so is one which makes it inappropriate to apply the same description to the other three. There is no state of *artha*, of *kāma*, or of *dharma* which a man may come to realize and rest in. Rather these terms are to be construed more subtly, perhaps as attitudes or orientations.

Artha is usually said to have to do with material prosperity. *Kāma* has specifically to do with sexual relations—and incidentally with aesthetic value. *Dharma* is said to concern the duties of one's station in life, the requirements laid down by one's involvement in family, caste, or class. It is certainly not at all easy to see the underlying rationale for this list, and calling them "aims," that is, attitudes or orientations, may not seem in any very evident fashion to help us understand the notions better. Nevertheless, if we take time to explore the concepts, I think the reader will see why calling them attitudes is perhaps the least misleading way of assessing them.

Consider first the kind of attitude one tends to take toward material objects that one has to handle from day to day in the routine of making a living—say, toward the door to one's office. How concerned are we for the door? That is to say, do we identify ourselves with the door and care what happens to it? Yes, but minimally. We care that the door opens and shuts when operated. We may care what it looks like, that it not have an unpleasant appearance. Beyond that, it would not occur to most of us to have a concern for the door. It is, therefore, an object of minimal concern for us, and it is the attitude of minimal concern which is the attitude marked out by the word *artha*.

Now such an attitude can, in principle, be taken toward any item in the world. Most people take an attitude of minimal concern toward some animals, or even all animals. Most of us take this attitude toward some people, reserving our attitudes of personal concern for relatives and close friends. A few people have no friends, and treat their relatives impersonally. There have even been cases where we might be justified in suspecting that a person has taken an attitude of minimal concern toward himself, although this has an air of paradox about it; nevertheless, one who kills himself in a fit of despair may well be thought to have, at least at the suicidal moment, considered himself as of minimal concern. When the Indians have said that *artha* has to do with material prosperity, I think they have had in mind that those things which are manipulated in making a living, including in some cases people or even oneself, that *those* things are typically things toward which we take an attitude of minimal concern. But it is the kind of attitude that the Indians were interested in, and the identification of this attitude by reference to material prosperity was merely a convenient way of marking down contexts in which this attitude is most frequently taken. It is not a defining characteristic of the attitude of minimal concern, *artha*, that it be directed toward those things with which we come in contact in our search for material prosperity, although it is in fact true that we tend to see this attitude exhibited frequently toward those things.

To take the *artha* attitude toward things is to exhibit maximal attachment coupled with minimal concern. With specific relation to the attitude toward money, the case that comes to mind is the seeker after wealth, the concupiscent miser or spendthrift. Such a person sees money as an end in itself or a means of gathering more material goods, of displaying more things' dependence on himself. He aggrandizes the things he desires, caring little for how their wishes or interests (if any) are served in so doing. His attachment is measured by his dependence on money; without it he views his own worth as negligible. As a result, but without his seeing it, he is completely dependent on things other than himself. If his wealth should perish, he surely would also be doomed.

THE ATTITUDE one tends to take toward those to whom one is passionately attracted has a very distinct feeling from the attitude of minimal concern. Again, one may take this attitude of *kāma* or passionate concern toward anything in the world. The little girl who hugs her teddy bear takes it toward an inanimate object, while the grown man who loves a woman passionately takes it toward an animate one. Again, it is possible to take this attitude toward oneself, as may be the case in narcissism. The mark of this attitude is an element of possessiveness in the concern one has for the object of one's attentions, cou-

pled with a partial identification of oneself with that object.

In the *kāma* relationship one depends upon the object that is loved and therefore guards it jealously, restricting if need be its habits or wishes (if it has them) in one's own interest. This is also true, to be sure, of the *artha* attitude, for one who takes an attitude of minimal concern toward a thing may guard it jealously and not hesitate to restrict its habits or wishes. But in the *kāma* attitude, unlike the *artha* one, there is a feeling that at some moments, at any rate, one is enlarged by becoming one with the loved one or thing. The little girl projects herself as the teddy bear and talks and feels on its behalf and receives satisfaction therefrom, as does the successful lover who in making love projects himself as the beloved and acts, for the moment, as if he were she or the two were one. Again it is paradoxical to speak of the narcissist projecting himself into himself; this seems to involve two uses of the word "self," for the self that projects is distinct from the self into which it is projected. But the paradox will be recognized, I think, as reflecting something essentially true. As before, when the Indians describe *kāma* in terms of sexual relations they do not mean to restrict the operation of this attitude to just those objects with which one can come into a sexual relationship, but are rather pointing to sexual relationships as typically involving instances of the taking of this kind of attitude.

To take the *kāma* attitude, then, is to exhibit a still strong attachment coupled with a growing but weak concern. Applying this to money, one can see that the lover sees money as a means of winning the affection of loved ones. But real love, one may well respond, requires maximal concern, unqualified commitment to an-

other's welfare. To put a price on sharing must surely qualify in some fundamental way that extent of one's concern for one's beloved. Maximal concern, in turn, is a state free from dependence—on money, on the beloved's affection, on anything else but one's pure concern for others. So the *kāma* attitude, though more admirable than that of the life of dependence and attachment characteristic of the *artha*-oriented person, falls short of true or real love insofar as it constitutes a claim on the other and thus a dependence on him or her.

W̲HEN WE TURN to the *dharma* notion, we find again that it is an attitude to which the term has primary reference. It is an attitude of concern greater than that involved in *artha* or *kāma*. In taking the *dharma* attitude, one treats as oneself things commonly thought of as other than oneself; not, however, in the way that the *kāma*-oriented man treats himself—in a spirit of passion and possessiveness—but rather in a spirit of respect. That is, not only do we in the *dharma* orientation project ourselves into others, but we do so with a certain conception of ourselves which precludes either using others or depending on them as the *artha*- or *kāma*-oriented person does. It is customary to render *dharma* as "duty" in English translations of Sanskrit works dealing with the "aims of life." This rendering has the merit that it suggests, following Kant's use of the word "duty," the crucial aspect of respect for the habits and wishes of others. It has the drawback, however, that it suggests to many people a rather stiff, perhaps even harsh, attitude, from which one tends to withdraw to something halfway between possessive love and "righteous" minimal concern. As a result, there is a supposed irreconcilability between an ethics of passionate concern for others and an ethics of duty which operates independently of such concern. But we need not wish this difficulty upon Indian thought; *dharma* does not mean "duty" in any sense of lack of concern for others— quite the opposite. The attitude of *dharma* is an attitude of concern for others as a fundamental extension of oneself. To see more clearly what this means we do well once more to appeal to examples.

Like the others, the *dharma* attitude can be taken toward anything, but the Indians mention certain things in the world as being more frequently its object. The attitude of a wise and loving mother toward her child is a case in point: The mother is concerned for her child as a fundamental and (within this orientation) indistinguishable part of herself. Her own self is enlarged by this identification. But since she respects herself—*i.e.*, she respects her own habits and wishes, unlike one who takes an *artha* or *kāma*

attitude to herself—she must also respect her child's habits and wishes, so that she neither uses the child for her own devices nor does she depend on the child in the way the passionate lover depends on his beloved. This is because the child is part of herself within this orientation. And generally speaking, since Indians have a strong feeling for the sacredness of family ties, relations of parent to child and grandchild, of child to parent and grandparent, are primary areas where one may expect to find the dharmic attitude exemplified. Not far removed, too, are the relations between oneself and the rest of one's clan, or caste, or class. Here, too, one may hope to find *dharma* instances, although all too frequently—especially in this day and age, so degenerate according to classical Indian tradition—one will be disappointed.

This *dharma* attitude involves weakening attachment and strengthening concern. Such a one sees money as a means of helping others to gain ends one views as worthwhile for them. He is willing, indeed eager, to part with his money to help others, provided he views their aims as right and their cause as just. He is able to love, showing concern for others, though this love is still tempered by judgments about relative worth, relative to *his* ends. These claims on him are, though, when he is being honest with himself, subject to doubt and suspicion—and when he is not so honest, the claims may be met with defensive insistence coupled with a refusal to succumb to the doubts and suspicions. Still, the dharmic person is, as has been frequently recognized by psychologists, less at the mercy of unwanted developments, more able to bear up under strain and hardship. His money is not everything to him, as it is to the miser, nor is it something he depends on for everything he holds dear, as it may be for the passionate lover who needs money to keep the attention of his beloved. Loss of such a one's money may (though unfortunately it may not) be the catalyst to propel him forward to the state of liberation, the fourth state which I am about to discuss.

THE ROUTE to superior control, to the fourth and most worthwhile kind of attitude, *mokṣa* or complete freedom, lies in the mastery of attitudes of greater and greater concern coupled with less and less attachment or possessiveness. In fact, the fourth orientation is well understood by extrapolating from this route. In moving from *artha* to *kāma*, we move from lack of concern to concern, and from more attachment to less. *Mokṣa* or freedom is the perfection of this growth. When one attains freedom, one is both not at the mercy of what is not oneself (that is to say, one is free from restrictions initiated by the not-self), and one is also free to anticipate and control anything to which one turns one's efforts, since the whole world is considered as one's self in this orientation. The *freedom-from* corresponds to one's lack of attachment, and the *freedom-to*, to one's universal concern.

The *mokṣa* attitude is one of maximal concern and minimal attachment. One is not attached to money or to making it; one is not at the mercy of any need for money, one is not dependent on having it. That is to say, then, that one is free to show universal concern for others. One's own needs are completely subjected so as to leave room for the needs of others, to behave so as to strengthen their concern and lessen their attachment. Indian myths and histories are full of examples illustrating behavior reflecting this combination of attitudes. Indian traditions covering the expectations about

the behavior of holy men underscore their conception of such a one's nature. As examples one may consult the *Jātaka* stories of the Buddha's former births, stories which range over the gamut of stages of his lives but include several in which he displays undemanding sympathy and understanding of others coupled with complete sacrifice for their betterment.

It is, of course, far too easy to misrepresent this ultimate stage of spiritual advancement. Nonliberated folk, among other shortcomings, are understandably incapable of accurately distinguishing the really liberated person from the consciously fake or unconsciously self-deluded display of some of the presumed features of liberation. The notion of liberation has the capacity to hoodwink the best-intentioned sympathizers, and people uncompelled by any vision of spiritual perfection find this a mark of the shortcomings of this entire way of viewing human aims. It is probable that current movements in India are undermining the classical theory of the aims of life and its implication of complete freedom—minimal attachment with maximal concern—as being the fundamental aim and purpose of life. Yet that way of thinking about human purposes still sways some of the ordinary folk in Indian villages, and in the now-distant past has moved ancient wise men to produce some of the most stupendously rich literature on human life and its ultimate purpose that has ever been composed. Perhaps it is not *passé* even now. ●

Adapted by the author, from Karl H. Potter, *Presuppositions of India's Philosophies* (Englewood Cliffs, N.J.: Prentice-Hall, Inc., 1963; Westport, Conn.: Greenwood Press, Inc., 1972, 1976), pp. 5-10.

EPICYCLE

A Case of Gold/Kurdish

A MAN WAS SITTING on a rock, overlooking the ocean. Next to him was a case filled with gold coins. The man picked up a coin, studied it, and then threw it into the ocean. He picked out another coin, turned it over in his hand, and flung it into the sea. A *q'alan-daar** was passing by, and stopped to look at this unusual performance. After seeing a good amount of gold disappear into the sea, he approached the man and said, "What is this?"

"A case of gold," the man answered.

"But what are you doing?" he asked.

"I'm throwing it into the ocean."

"What do you want to do that for?"

"I am practicing non-attachment," he replied, calmly.

"Then why don't you just dump them all in at once?" asked the *q'alan-daar*.

"Oh, no," he said. "This attachment I have, it needs to be struggled with a hundred thousand times." ●

—*Retold by Manocher Movlai*

* *Q'alan-daar*— A man who has reached a certain level of understanding of life.

From *Walking Into The Sun: Stories My Grandfather Told*, collected by Jon Schreiber. © 1991 by California Health Publications, 6201 Florio Street, Oakland, CA 94618.

CURRENTS . . .

EXHIBITS

The Schmidt Bingham Gallery, New York City (212-888-1122), announces a travelling exhibit of the paintings of Morris Graves, **Vessels of Transformation, 1934-1986**, which will be on view at the Art Museum of Southeast Texas, Beaumont, Texas, from February 8 through March 27; and at the Rahr West Museum, Manitowac, Wisconsin, from April 14 through May 22. The exhibit includes thirty-six paintings and drawings of the artist's visionary period of the 1930s and his floral images of the '50s through the '80s.

The Asia Society, New York City (212-288-6400), presents **Romance of the Taj Mahal**, through March 17. The exhibit includes seventeenth-century art from the reign of Shah Jahan, the Mogul emperor who created the Taj Mahal, and later works inspired by the monument itself.

The Arthur M. Sackler Gallery, Washington, D.C. (202-357-2700), will present **Paper and Clay from Modern Japan** through March 17. The twentieth-century paper and ceramic objects include works of several potters designated as Japan's "Living National Treasures."

The Santa Barbara Museum of Art, Santa Barbara, California (805-963-4364), will present **Later Chinese and Japanese Bronzes** from February 2 to April 21, an exhibit featuring sixteenth to twentieth-century Chinese and Japanese vessels and mirrors. The museum also announces **Comrades and Cameras: Photographs from the Baltic and Soviet Republics**, from February 9 to April 28. Included will be 135 works by more than 50 photographers documenting the pre-dawn of glasnost.

The Museum of Fine Arts, Houston, Texas will present an ongoing exhibit entitled **African Gold: Selections from the Glassell Collection** beginning March 10. The exhibit includes objects created by the Akan peoples of the Ivory Coast and Ghana as well as the Fulani of Mali and the Swahili of Kenya.

SYMPOSIA

Cauldron Productions, New York City (212-245-0445), will present **Death— The Secret of Life** at St. Peter's Lutheran Church, New York City, on Saturday, February 9, from 8:30 A.M. to 10 P.M. The one-day symposium will include lectures and programs of storytelling, theater, film, and dance featuring Wendy Doniger, Robert Thurman, Robert Venables, Astad Deboo, Steven Fadden, and others.

The Kairos Foundation (London) and The Dallas Institute of Humanities and Culture (214-327-5688) will present **Art and the Sacred**, an exchange among representatives of a wide range of international sacred traditions, in the St. Francis Auditorium of the Museum of Fine Arts, Santa Fe, New Mexico, from March 21 to 24. Speakers will include the Very Reverend James Morton, Keith Critchlow, Yoshikaza Iwamoto, and Therese Schroeder-Sheker. There will also be exhibitions, recitals, and concerts in a cross-cultural exchange of the sacred traditions in the arts.

The Mind/Body Medical Institute and Tibet House New York announce **Mind Science: A Dialogue Between East and West** with His Holiness the Dalai Lama on the weekend of March 23 and 24, at the Deaconess Hospital of the Harvard Medical School, Cambridge, Massachusetts.

MUSIC & DANCE

The World Music Institute, New York City (212-545-7536), will present the **Satyanarayan Charka Kathak Dance Company** at Symphony Space, New York City, on Saturday, March 2 at 8:00 P.M.; and **Music of Persia** with Dariush Talai and Madjid Kiani at the Washington Square Church, New York City on Saturday, April 7 at 8:00 P.M.

Richard Wagner's **Ring of the Nibelung** cycle will be performed twice this summer at the Seattle Opera (206-443-4700), August 3 to 8 and August 11 to 16. The second part of the cycle, **Die Walküre**, will be in repertory at the Lyric Opera of Kansas City (816-471-4933) in April. Wagner's final opera, **Parsifal**, based on the legend of the Holy Grail, will join the repertory of New York City's Metropolitan Opera (212-362-6000) in March, with tenor Placido Domingo in the title role and soprano Jessye Norman as Kundry. The Wagner Society of New York (212-749-4561) will hold an all-day seminar on **Parsifal** on Saturday, March 30, at Christ Church in New York City.

EDUCATIONAL OPPORTUNITIES

Penn State University (814-863-2284), in affiliation with the Native American Indian Leadership Program, is offering master's degree fellowships for Native American students interested in special education training. The teacher training program is designed to prepare teachers to work effectively with mentally and physically handicapped children. Seminars focus on Native American Indian education. The deadline for submitting applications for the fall semester 1991, which begins in August, is April 15, 1991.

... & COMMENTS

PROBABLY NOWHERE is the symbolism of money more intriguingly quoted than in the Tarot, the history of which has been exhaustively chronicled in Stuart R. Kaplan's massive, three-volume **An Encyclopedia of Tarot** (Stamford, CT: U.S. Games Systems, Inc., 1978, Vol. I; 1986, Vol. II; and 1990, Vol. III). Kaplan sees himself more as a historian than an interpreter, and his presentation is admirably thorough and objective—objective almost above and beyond the call of duty, in fact, considering the consummate silliness of such entries as the Teddy Bear Tarot or the various New Age excursions which run the gamut from comic-book eroticism to esoteric mumbo jumbo.

(Continued on page 98)

For fifteen years PARABOLA has gathered distinguished writers and thinkers seeking to reach new understandings of the universal human themes as they are expressed in myth and legend, ancient symbol and sacred art, folklore and ritual. Through PARABOLA back issues, you can now share in this storehouse of knowledge and insight. PARABOLA readers treasure their back issues as a permanent part of their libraries, a source of ideas to re-read and reflect on. Each is a beautifully designed, sturdily bound paperback of 128 pages. Back issues are $8.00 per copy; orders of 12 or more, $7.00 per copy.

Please use the order form bound between these pages.

☐ VOL. I:1 **The Hero** Mircea Eliade, Barbara G. Myerhoff, Barre Toelken, P.L. Travers, Jacob Needleman, Edward Edinger, Minor White; Huston Smith interview.

☐ VOL. I:2 **Magic** Barbara G. Myerhoff, Daniel Noel, Robert Ellwood, Jacob Needleman, Victor Turner, Thomas Moore, Christmas Humphreys; Joseph Campbell interview.

☐ VOL. I:3 **Initiation** Sam Gill, Janwillem van de Wetering, Arthur Amiotte, Evelyn Eaton, Fernando Llosa Porras; Mircea Eliade interview.

☐ VOL. I:4 **Rites of Passage** Frederick Franck, James Wolfe, Ursula K. Le Guin, D.M. Dooling, Robert E. Meagher; William Irwin Thompson interview.

☐ VOL. II:1 **Death** P.L. Travers, Conrad Hyers, Isaac Bashevis Singer, David Steindl-Rast, William Doty, William Burke Jr.; Tibetan Lamas interview.

☐ VOL. II:2 **Creation** Sam Gill, P.L. Travers, David Rosenberg, David Johnson, Jane Yolen, John Fentress Gardner, Daniel Whitman; Zalman Schachter interview.

☐ VOL. II:3 **Cosmology** David Steindl-Rast, Ursula K. Le Guin, Lorel Desjardins, Elaine Jahner, Anne Bevan, Harry Remde; Lloyd Motz interview.

☐ VOL. II:4 **Relationships** Frederick Franck, Robert E. Meagher, Shems Friedlander, Lizelle Reymond, Jean Toomer, Barre Toelken, Jane Yolen; Diane Wolkstein interview.

☐ VOL. III:1 **Sacred Space** A.K. Coomaraswamy, Barbara Stoler Miller, Robert Lawlor, Irving Friedman, Pablo Neruda, Hélène Fleury; P.L. Travers and Michael Dames interview.

☐ VOL. III:2 **Sacrifice and Transformation** Annemarie Schimmel, Joseph Epes Brown, Robert A.F. Thurman, Minor White; Adin Steinsaltz interview.

☐ VOL. III:3 **Inner Alchemy** Mircea Eliade, D.M. Dooling, Harry Remde, Jacob Needleman, Elémire Zolla.

☐ VOL. III:4 **Androgyny** Elaine H. Pagels, Titus Burckhardt, Keith Critchlow, P.L. Travers, Barbara G. Myerhoff; Lobsang Lhalungpa interview.

☐ VOL. IV:1 **The Trickster** Emory Sekaquaptewa, Michel Waldberg, Lynda Sexson, Barbara Tedlock, P.L. Travers, David Leeming; Joseh Epes Brown interview.

☐ VOL. IV:2 **Sacred Dance** Elaine H. Pagels, Rosemary Jeanes, David P. McAllester, Fritjof Capra, Annemarie Schimmel; Peter Brook interview.

☐ VOL. IV:3 **The Child** Don Talayesva, Richard Lewis, Frederick Franck, Lynda Sexson, Lobsang Lhalungpa, art and stories by children.

☐ VOL. IV:4 **Storytelling and Education** Maria José Hobday, Thomas Buckley, James Hillman, education symposium.

☐ VOL. V:1 **The Old Ones** Keith Critchlow, Agnes Vanderburg, Lobsang Lhalungpa, Robert Bly, Gary Snyder; Deshung Rinpoche and Joseph Campbell interviews.

☐ VOL. V:2 **Music Sound Silence** Herbert Whone, Tomas Tranströmer, David A. Lavery, Tom Moore, David P. McAllester, Howard Schwartz, Robert Lawlor; Steve Reich interview.

☐ VOL. V:3 **Obstacles** Al Young, Abraham Menashe, David Steindl-Rast, Jonathan Omer-Man; Mohawk Chiefs and the Dalai Lama interviews; photo essay.

☐ VOL. V:4 **Woman** P.L. Travers, Helen M. Luke, Seonaid Robertson, Ursula K. Le Guin, Barbara Rohde, Joseph Campbell; Judy Swamp interview.

☐ VOL. VI:1 **Earth and Spirit** Peter Matthiessen, Peter Nabokov, Robert Bly, Paul Caponigro, P.L. Travers, John Kastan, D.M. Dooling; Firoze M. Kotwal interview.

☐ VOL. VI:2 **The Dream of Progress** Kathleen Raine, David Malouf, Dino Buzzati, Seyyed Hossein Nasr; Chinua Achebe and Jacob Needleman interviews.

☐ VOL. VI:3 **Mask and Metaphor** Terry Tafoya, Ray Zone, Demorest Davenport, Adin Steinsaltz, Henrich von Kleist; Peter Brook interview.

☐ VOL. VI:4 **Demons** J. Stephen Lansing, Maria Dermoût, Chinua Achebe, Susan Stern, Edwin Bernbaum, Dino Buzzati; Isaac Bashevis Singer interview.

☐ VOL. VII:1 **Sleep** Henri Tracol, Mircea Eliade, A.K. Ramanujan, Jonathan Omer-Man, P.L. Travers; Dr. Yeshi Dhonden and Joseph Campbell interviews.

☐ VOL. VII:2 **Dreams and Seeing** Ursula K. Le Guin, Arthur Amiotte, Wendy Doniger O'Flaherty, Elémire Zolla; conversation between Laurens van der Post and P.L. Travers.

☐ VOL. VII:3 **Ceremonies** Joseph Epes Brown, Robertson Davies, Doris Lessing, Francelia Butler, Frederick Franck, Barbara Nimri Aziz, P.L. Travers, David Abram, Joseph Bruchac, David Steindl-Rast.

(Continued from page 95)

Kaplan focuses primarily on the twenty-two cards of the Major Arcana, but the minor suits—including, of particular interest for this issue, Coins or Pentacles—are also sampled regularly, especially in decks like Rider-Waite, where the cards in the suits are depicted as fully drawn metaphorical scenes, not just the number of Coins (or Cups or Wands or Swords) indicated. Kaplan points out that Coins are related not only to money, but also to talents, commerce, wheels or cycles, charity, and the element of Earth. He quotes a sixteenth-century writer, Pietro Aretino, as saying that the suit represented "the sustenance and food of play" in the "hazardous game" of life.

Also of particular interest is the Hanged Man from the Gringonneur deck, shown here, designed for King Charles VI of France. The figure seems to have reference to both Judas, weighted down by his thirty pieces of silver, and the apostle Peter, who was crucified upside down and who is related to the astrological sign of Pisces (the twelfth sign of the zodiac, just as the Hanged Man is the twelfth card of the Tarot). But the card represents betrayal and denial—or static imprisonment—only on the surface. Its deeper meaning is related to sacrifice and transformation, for the figure, with a curiously relaxed expression on his face, has his feet not on the ground (not rooted in the materialism of the Earth), but in the sky: free, in touch with the Divine, open to involution—like money itself, in its ideal, original state.

■ ■ ■

THE QUESTION of real value or worth, of the relationship between the levels of the divine and the ordinary, is central to Arnold Schonberg's **Moses und Aron**, which entered the repertory of the New York City Opera last fall. The production effectively illuminates the central symbols that Schoenberg employs: the Burning Bush, as the pure and essential Word of God; and the Golden Calf, as the perverse materialization of that impulse to make it "accessible" to the general public. Linked to the two symbols are the two title characters, based on the Biblical Moses and Aaron, the former who only speaks in the opera (instead of singing), doing so with great difficulty, whereas Aaron sings melodiously and with great ease, turning Moses' struggle to express the ineffable into an oversimplified, popularized message for the masses. "Do not expect the form before the idea," Moses pleads in vain, as Aaron rushes on, like a wheeler-dealer promoter or press agent, even turning the Staff of Law into a snake as a cheap trick in front of Pharoah's priests.

—R.B.

FULL CIRCLE

A Readers' Forum

THANK YOU for Wendell Berry's essay on "The Body and the Machine" (XV:3), and his reiteration of the basic question "What is enough?" I'm inspired by the example of Rikyu, the tea instructor to the Shogun, Hideyoshi. At the time the shogun had a tea room in which every object was either gold or covered in brocade or rare and old. Rikyu cultivated a *wabi* life of simplicity and told Hideyoshi that to pay the greatest honor to his guest would be to make tea with his own hands and serve his guest, in a hut resembling a hut in the mountains, far from the world's concerns.

—*Lewis Headrick*
San Francisco, California

I SAW THE Bill Moyers' interview with Robert Bly (reviewed in PARABOLA XV:3) when it appeared on PBS sometime back and found two things in it that bothered me.

I remember Robert Bly as being a poet of powerful images and strength of verse. The poetry that he recited on the program seemed to have lost this strength. Perhaps his search for his father is diluting his poetry.

Secondly, I am still at a loss for the reason for the "men's movement." Some years ago it was understandable, though regrettable, that blacks and women felt it necessary to go into a phase of separatism. After all, they had been disempowered and needed to see that they did possess worth as human beings. However, white males have always been in power in our society. They don't need the separatist

phase. If they have discovered that something is wrong, then they need to reintegrate themselves into the human race, not into the male race. Men need to become human again.

—*Don Redmond*
DeSoto, Illinois

IN LOOKING over the Winter issue of PARABOLA focused on the theme "Triad," I was disappointed to find no mention of Patrizia Norelli-Bachelet, preeminent cosmologist of today. While it is always helpful to provide historical background from known sources on any subject, surely it is equally valuable to present an evolving new vision as well.

Patrizia Norelli-Bachelet's work with number qualifies her to say some very interesting things on the nature of Three. Because her work is not speculative but rooted in the reality of today, she is able to offer living examples of her observations and number comes alive. The Earth's position in the Solar System gives us a primary clue about Three. Patrizia Norelli-Bachelet writes in *The Gnostic Circle*:

. . . if we speak of the Earth as the planet upon which the Trinity incarnates for which it is brought into the play of crea-
(Continued on page 104)

PARABOLA BOOKS

I BECOME PART OF IT
Sacred Dimensions in Native American Life
Edited by D.M. Dooling & Paul Jordan-Smith
Introduction by Joseph Bruchac

An anthology of compelling articles by Native Americans such as Arthur Amiotte, Vine Deloria, Oren Lyons, and Emory Sekaquaptewa, and Native American scholars Joseph Epes Brown, Sam Gill, Barbara Tedlock, and Barre Toelken, interspersed with traditional Native stories and contemporary art.

"An important and unusually accessible contribution."— Kirkus

Softcover; 304 pages, $14.95

DARK WOOD to WHITE ROSE
Journey and Transformation in Dante's Divine Comedy
by Helen M. Luke

In this provocative study, Helen Luke takes the reader on an intense spiritual and symbolic journey of exploration into each of Dante's poetic images. Richly illustrated with reproductions in black and white of medieval manuscript paintings inspired by Dante's masterpiece.

"A proud book with sensitive prose . . . The mystical journey toward wholeness as viewed from a Jungian perspective."
— The Book Reader

Hardcover; 224 pages; 44 illustrations; $19.95

THE SONS OF THE WIND
The Sacred Stories of the Lakota
Edited by D.M. Dooling

The first narrative presentation of the mythology of the Lakota Sioux based on the Walker manuscripts. The stories are centered around the "Sacred Beings," the personalizations of the forces and spiritual qualities that underpin the Lakota universe.

"An eloquent and accessible rendering of the mighty creation myth of the Lakota people, which ranks among the most stirring in earth literature."
— Peter Matthiessen

Paperback; 168 pages; $8.95

THE GARDEN BEHIND THE MOON
A Real Story of the Moon Angel
Written and Illustrated by Howard Pyle
"Best Children's Storybook, 1989"
— Publishers Marketing Assoc.

The long-out-of print children's fantasy tale of a boy drawn by the mysterious Moon Angel into following the moon path to its source. Meant to be read aloud, with over 30 original illustrations and woodcuts.

"Absolutely charming! Like all great mythic stories it delves into all the deepest and darkest questions which the human spirit has asked throughout the ages ."
— Madeleine L'Engle

Hardcover; 192 pp.; over 30 illustrations; $14.95.

AND THERE WAS LIGHT
The Autobiography of Jacques Lusseyran

This previously out-of-print classic is a gripping, heroic story by a man who transcended both blindness and imprisonment in a Nazi concentration camp to develop a unique understanding and love of life. It is a story of blindness and seeing, and how they exist together to create an inner light.

"This is a magical book, the kind that becomes a classic, passed along between friends."   — The Baltimore Sun

Paperback; 320 pages; $10.95

OLD AGE
by Helen M. Luke

In this collection of essays, Helen Luke, Jungian counselor and Director of the Apple Farm Community, gives us her challenging thoughts on the transformative mysteries of old age and suffering. "Growing old" is explored as a journey of renewal and new challenges as the Self moves toward completion and wholeness.

"I know of no woman of our time who has made a richer and more significant contribution, in a truly contemporary way, to man's understanding of himself. Her essays on old age are among the finest she has ever done."
— Sir Laurens van der Post

Hardcover; 128 pages; $12.95

A WAY OF WORKING
The Spiritual Dimension of Craft
Edited by D.M. Dooling

In this enriching collection of eleven interrelated essays by a group of authors writing not primarily as individuals but as members of a community, the ancient relationship of art, order, and craft is explored. Craft is considered as a "sort of ark" for the transmission of real knowledge about being, and about our deep creative aspirations.

"This book is not only for any serious craftsman but for everyone who wants to make sense of his one and only life."
— Louise March, Founder and Director, Rochester Folk Art Guild

Paperback; 142 pages; $8.95

All Parabola Books have sewn bindings and are printed on acid-free paper.

Please use the order form bound between pages 96 and 97.

THE BESTIARY OF CHRIST
by Louis Charbonneau-Lassay
Translated by D.M. Dooling

*"The characteristics of the hedgehog as the allegorical image
of Satan, and those of the sheep as the emblem of the Savior, are consequently
in complete opposition: the sheep is soft to the touch, the hedgehog harsh
and painful; the sheep is docile and quiet, and the hedgehog on the contrary
curls up into a bristling ball as soon as anyone approaches it."*

Just before the outbreak of the Second World War in Europe, a little-known Roman Catholic scholar living in the east of France published a compendium of animal symbolism that ranks with the greatest of the classical and medieval bestiaries. Louis Charbonneau-Lassay's *Le Bestiaire du Christ* (*The Bestiary of Christ*) was a tour de force that brought together the findings of a lifetime of scholarship in religious symbols gleaned from sources as diverse as ancient Egypt, classical Greece and Rome, early and medieval Christianity, the Kabbalah, Gnosticism, and various spiritual schools of the Near and Far East.

Bringing together various schools of esoteric wisdom with Catholic thought and the folk legends of the French countryside around Loudun, where he lived and died, Charbonneau-Lassay created a stirring and lively account of the rich—and often contradictory—metaphorical meanings of real and imaginary animals. His free-wheeling and exhaustive compilation ran to over a thousand pages and included a thousand original woodcuts.

This new edition—the first ever in English—has been translated and abridged by D.M. Dooling, the founding editor of PARABOLA, the Magazine of Myth and Tradition. As is stated in the introduction, "Evidently, not many people have had a chance to read it; yet the rumor of it has spread slowly, almost secretly, as if the magic of the old symbols, reinvested through the love and sensitivity of the author with the power of their ancient meanings, traveled on some unknown wavelengths to reach out, fifty-odd years later, to another audience. Bits and traces of the original book were found here and there; I knew of it long before I held a copy in my hands—and after that, slowly but inevitably, it became necessary to pass it on to others."

496 pages, hardcover; 400 illustrations $29.95
ISBN 0-930407-18-0
Please use the order form bound between pages 96 and 97.

(Continued from page 100)
tion, this can be further verified by the fact that the Earth occupies the 3rd point, or the 3rd orbit; hence, the Law of Three, from the Divine Trinity down to the individual's psychic being and to the simple atom, is always the supreme pattern and rhythm of the planet.

—*Barbara White*
Stone Ridge, New York

THE PARABOLA issue "Time and Presence" (XV:1) precipitated a noteworthy series of coincidences and spiritual insights for me.

The Kabbalah study group in which I participate was discussing the concept of "connections," the exciting moments when interpersonal or solitary consciousness is transcended and new levels of awareness are reached. The day of our meeting happened to coincide with the one-year anniversary of my father's death. In place of the traditional candle which the group lights at each meeting, that day I kindled a Yahrzeit candle, the Jewish memorial symbol, and set it in the center of the circle.

The night before I had read Jon Swan's enchanting poem "Father Father Son and Son." I shared the poem with the group that morning, following which our discussion shifted to topics of reincarnation and the post-Holocaust legacy of Jewish spiritual values.

M.C. Huntoon's article "Another Dimension" projected a pencil passing through multi-dimensional embodiments, perceived in each Planeworld as a cross-section of the invisible whole. The hypothesis of a string projection of multiple dimensions bridged by a mechanism which our limited senses cannot conceive struck a personal note.

Presence, like the afterglow of a sudden illumination, can transcend dimensional limitations. Like sunlight crossing the chill of space, leaving warmth to the body that can contain warmth, a symbolic candle served as an acute reminder that a single soul of the millions set free in genocidal madness have not departed but have been shifted to another dimension, still immanent but not manifest in traditional means.

As other articles in the issue indicate, time is the administrator of presence, providing physical access to only the middle third of the past-present-future continuum. The other doors are opened by spirit, not by hand.

—*Marvin Schwartz*
Conway, Arkansas

FIRST, I would like to thank you for your excellent publication. It is a real pleasure to read each issue—usually cover to cover!

During the days when I was immersed in your Winter issue on Hospitality (XV:4), our hermit—who lives in an old trailer on the monastery's property—gave our abbot a note which was later posted on the community bulletin board:

Last night I had an unexpected visitor. At about 10:40 I heard a loud thump. I awoke to see a light outside the chapel window and heard another loud thump. I called out, "Come in!" and lit a candle. Whoever it was, however, had fled—the window and molding were broken.

I read this account—this "Come in!" and, in the vernacular of the *Sayings of the Desert Christians*, I "went away greatly edified." His response minds me of St. Benedict's exhortation in his *Rule*: "All guests who present themselves are to be welcomed as Christ."

Benedicite Dominus.

—*Br. William Lawder, OCSO*
Lafayette, Oregon

Personal transformation—
It's just a matter of knowing where to look.

In these essays, handbooks, and personal narratives,
outstanding authors explore the power of myth, ritual, and spirituality
in many cultures — and in our own lives.

MYTHOS
THE NEW SERIES IN WORLD MYTHOLOGY

Classic books in new paperback editions

The Hero with a Thousand Faces

Joseph Campbell

Hero is a genuine classic. More than 500,000 readers have already enjoyed this compelling and fascinating study of the myths of the hero—more than 200,000 new readers in the last two years alone.

It's time to discover (or rediscover) **Hero** for yourself. Walk through a few centuries with Joseph Campbell as your guide. It is a very rewarding journey.
Bollingen Series ✠

Now in paper: $9.95 ISBN 0-691-01784-0
Not available from Princeton in the United Kingdom

In Quest of the Hero

Otto Rank, Lord Raglan, and Alan Dundes
With a new introduction by Robert A. Segal

Three essential works are brought together for the first time: Rank's *Myth of the Birth of the Hero*; key selections from Raglan's *The Hero*; and Dundes's "The Hero Pattern and the Life of Jesus." Examined are the patterns found in the lore surrounding historical or legendary figures like Gilgamesh, Oedipus, Odysseus, Perseus, Heracles, Romulus, Arthur, and Buddha. The perfect companion to Campbell's *Hero*.

Now in paper: $10.95 ISBN 0-691-02062-0

The Wisdom of the Serpent
The Myths of Death, Rebirth, and Resurrection

Joseph L. Henderson and Maud Oakes

"...offers crucial insights on the meaning of death symbolism (and its inevitably accompanying rebirth and resurrection symbolism) as part of the great theme of initiation, of which [Henderson] is the world's foremost psychological interpreter. This material is really the next step after the hero myth that Joseph Campbell has made so popular...."—*John Beebe*

Now in paper: $10.95 ISBN 0-691-02064-7

Avicenna and the Visionary Recital

Henry Corbin
Translated from the French by Willard R. Trask

A translation and analysis of three mystical "recitals" of the great eleventh-century Persian philosopher and physician Avicenna, author of over 100 works on theology, logic, medicine, and mathematics.
Bollingen Series ✠

Now in paper: $16.95 ISBN 0-691-01893-6

BOOK REVIEWS

What Are People For?
By Wendell Berry. San Francisco: North Point Press, 1990. Pp. 220. $19.95, cloth, $9.95, paper.

As with his poetry and fiction, Wendell Berry's essays and reviews are nurtured by all of his work, which would include not only his labor as a Kentucky farmer but also as a husband, father, and neighbor. "I now live in my subject," he explains simply in the first essay in this new collection. "My subject is my place in the world, and I live in my place." If this is so, the book reminds us, then it becomes inappropriate, immoral even, to make the subject into an object, or even to think any longer in terms of objects. "I am a part of all I see," Whitman exclaimed; while less exuberant, Berry could say the same.

Many of Berry's essays are structured to mirror their own central point, that everything grows from the local. "The Work of Local Culture," for instance, begins "For many years, my walks have taken me down an old fencerow in a wooded hollow on what was once my grandfather's farm. A battered galvanized bucket is hanging on a fence post. . . ." We hear the old bucket's story, learn of the "slow work of growth and decay" which "has by now produced in the bottom of the bucket several inches of black humus." The bucket collects leaves and stories; it is both literal and, like Berry's writing, "irresistibly metaphorical."

What Are People For? the title asks. The bucket, Berry answers, "is doing in a passive way what a human community must do actively and thoughtfully." Human culture is a kind of compost that preserves through proper use and transformation. Like an old bucket that collects leaves that eventually become new soil, "a human community too, must collect leaves and stories, and turn them to account. It must build soil, and build that memory of itself—in lore and story and song—that will be its culture."

Berry's own voice as a writer would not exist without the family, the land (both farm and wilderness), and the larger community to which he is attached. If it is his individual character we meet in these pages, that too is always fixed in its locale. Style, as he demonstrates here, results from deeply rooted choices. And if his style is in part one of accumulation— building ordered, crafted sentences and paragraphs through (for example) a judicious handling of parallelisms—it accumulates the way soil does. It works well with the materials at hand, turning everything to good use.

To lose the sense of living and working within one's subject is to risk a kind of dislocation which makes possible the exploitation of land and people. It is to see both as "raw material." To do so makes not only for poor farming, but also for poor ethics *and* aesthetics. Berry exemplifies this in a comparison of Hemingway's great fishing story, "The Big Two-Hearted River" and Norman Maclean's "A River Runs Through It." Hemingway's story, he concludes, "seems an art determined by its style. This style, like a victorious general, imposes its terms on its subject." Neat, safe, and controlled, like Nick Adams in his tent, "it deals with what it does not understand by leaving it out." In con-

trast, Maclean's story uses fishing as "emblematic of all that makes us companions with one another, joins us to nature, and joins the generations together. This is the connective power of culture." It is, then, "a style vulnerable to bewilderment, mystery, and tragedy—and a style, therefore, that is open to grace."

A tension exists in our culture between the need for the wild and the ordered; this tension is reflected in much of the writing that Berry celebrates in the second half of this volume (that of Edward Abbey and Wallace Stegner, among others). In his insightful analysis of *Huckleberry Finn*, Berry comments that Huck flees a decent home as if there could be no choice between "either a deadly 'civilization' of piety and violence or an escape into some 'Territory' where we may remain free of adulthood and community obligation." Our country's culture remains suspended in a kind of protracted adolescence, he suggests, that fails to see that real life and grace are found in the daily responsibilities at home, in the community, and on the land, and that these in turn must stay in touch with the wild in order to remain honest.

I'm reminded here, and through this new volume, of the relationship between the lawyer Wheeler Catlett and his old friend Burley Coulter in Berry's story "The Wild Birds." It is a friendship of contrasts: "If Burley has walked the marginal daylight of their world, crossing often between the open fields and the dark woods, faithful to the wayward routes . . . Wheeler's fidelity has been given to the human homesteads and neighborhoods and the known ways that preserve them." But in this story Wheeler must go beyond the safe places he prefers in order to met Burley on the margins of things, in the wild places. He must honor the otherness in his friend—and through him honor the wildness in all things, including himself. He must learn, as Berry has learned, that the cultivated human order depends upon our connection to the untamed.

Like all Wendell Berry's books, *What Are People For?* is a book about cultivation. Because of this it is also a book about relationships and responsibility, about character and community, about dignity and discipline. It is appropriate, I think, that the book's final word should be "marriage," which so often surfaces in Berry's work as a gathering point for much of what he affirms: a kind of fidelity to life, to the land, to one's neighbors, friends— and to the wild places both within and without us.

I continue to read Wendell Berry not with the expectation of learning something new but with the certainty of being reminded of something very old. I read him for what he calls forth in me. I see again the importance of care taken over small things, including the great discipline that words and sentences bring (including the physical making of a book: praise be to North Point!). And I see again the joy that comes from such care.

Perhaps what strikes deepest in my reading of this book is the reminder he gives of the joy in good, honest work, including the hard work of growing up. This is clear-eyed realism in its truest spiritual sense: he does not blink from the fact that one day his own death, like that of his father before him, will enrich the soil he has farmed. Yet still he takes delight in being part of the collective memory of the tribe, part of the ongoing human story.

—*Douglas Thorpe*

Douglas Thorpe is the author of the forthcoming A New Earth *from Catholic University of America Press.*

DEPTH PSYCHOLOGY OVERSEAS

Announcing Pacifica Graduate Institute's M.A. Overseas Study Program

Recognizing the importance of the landscape as a primary architect of the human experience of the psyche, Pacifica Graduate Institute now offers its Master of Arts Degree in Counseling Psychology in a program that includes overseas study.

The Pacifica Overseas Program combines classes at the Institute's Santa Barbara campus with summer sessions that alternate annually between the Hawaiian Islands and selected European locations. In these settings academic coursework is combined with the on-site study of arts, literature, and mythology in the lands where they originated. It is an experience designed to enrich the discovery and exploration of psyche-centered psychotherapy.

Pacifica's M.A. Degree Program emphasizes the traditions of Depth Psychology pioneered by C.G. Jung.

Classes at the Santa Barbara campus are in session once each month over a three-day weekend.

Upon completion of curriculum and internship requirements, graduates are eligible to sit for the California Marriage, Family, and Child Counseling License exam.

PACIFICA
GRADUATE INSTITUTE

For more information:
249 Lambert Road
Carpinteria, California 93013
(805) 969-3626

Crow and Weasel

By Barry Lopez. San Francisco: North
Point Press, 1990. Pp. 64. $16.95,
cloth.

Badger informs the animal
heroes of Barry Lopez's illustrated fable *Crow and Weasel* that
"the stories people tell have a way of
taking care of them. If stories come to
you, care for them. And learn to give
them away when they are needed."

The author spent ten years with
the telling of this vision quest. His
tale is set in "myth time," when both
animals and men wore clothing, rode
horses, and talked. As in a classic Native American story, the outer and inner landscapes are closely linked in the
telling. The young Crow and Weasel
leave their tribal home and venture out
farther than any of their people have
gone before, searching for some sign,
some marvel to give meaning to their
life. As they cross the land, they begin
to understand their connection to it.
But it is only when they meet the flat-
faced people, the Inuit, for the first
time, sharing food and stories with
them, that they truly know what their
quest has been about.

Lopez's deep, abiding concern for
the environment has been demonstrated before—in his National-Book-
Award-winning *Arctic Dreams* and in
Of Wolves and Men, as well as in his
strong, elegant essays. While I applaud
his efforts at storytelling in *Crow and
Weasel*, the result is more often non-
fiction than myth. His attention is on
tangibles—clothing, food—rather than
on the growth of character or the sense
of the Otherworld that touches many
Native American vision tales.

The beautiful moments in this
little book are descriptive: the details
of lighting the medicine pipe, with its
mixture of bearberry leaves and dried
willow; Weasel's list of what makes a
good horse; the careful, studious way
that Crow and Weasel dress their unknown prey:

Crow and Weasel skinned the caribou with
care, examining the meat and bones and
sinews closely, and the hide. The animal
had a kind of hair they had not seen before, very warm looking. And, they
thought, large feet for such a creature. Its
antlers were in velvet, and strange as well,
neither like a moose's antlers nor an elk's
but somewhere in between. Far into the
evening Crow and Weasel inspected the
animal, looking at the way the bones fit
together, where the heart was, and at the
shape and size of the lungs and the liver.
They looked at the contents of the stomach. They wanted to learn the animal.

But descriptions cannot carry a
mythic tale; I kept wishing for a storyteller to appear. I wish, too, that the
designer had been more familiar with
books in which text and illustration
mesh. Lovely as Tom Pohrt's pictures
are, occasional poor placement destroys their power. Twice, pictures and
text are out of synchronization—first,
when the boys are following a flicker
but the illustration is of a grouse
which is not introduced until the next
page; second, when Crow and his
horse fall through the ice, the picture
coming two pages after we know that
he has been rescued. Peculiar, too, is
that all the male animal characters—
Crow, Weasel, Grizzly Bear, Mountain Lion, Mouse—are shown elab-
orately dressed in Plains Indian
clothes, but the one female character—
Badger—is not dressed at all.

Also, a sharp-eyed editor should
have noted inconsistencies in the text.
Weasel and Crow kill a caribou and
make pemmican, talking about how
they will have plenty of food, and yet
three pages later—in the time of the
story, the next day—they are "short on
food." Later we are told within two
paragraphs that "the sameness of the
land disoriented them . . ." and that
"the country was not really the same."

Perhaps one should simply fall
through the words into the story, ig-

noring problems with picture placement, textual inconsistencies, even the slowness of the telling. But if, as Lopez means us to believe—and as I *do* believe—we must care for our stories, then such rigid standards for stories in books must be maintained.

I enjoyed this book, wish it had been much better for all the years of effort, and will give it away to our local library where—with budget cuts—it is needed.

—*Jane Yolen*

Jane Yolen is the author of over one hundred books, including Favorite Folk Tales from Around the World *(Pantheon) and the recent* Sky Dogs.

The Book of J
Translated from the Hebrew by David Rosenberg, Interpreted by Harold Bloom. New York: Grove Weidenfeld, 1990. Pp. 340. $21.95, cloth.

ANYONE WHO TRIES to grasp the meaning of the Hebrew Bible, ignoring the corpus of traditional commentary, will quickly encounter a number of puzzling phenomena. The same story is often told twice in succession, in significantly different versions. God is sometimes pictured as godlike and exalted, but at other times as outrageously human. The documentary hypothesis, formulated by nineteenth-century German biblical scholars, addresses these problems by suggesting that the Bible does not speak in one voice but is actually a skilled blending of several earlier, distinct strands, the oldest of which is called J because its name for God is Yahweh (Jehovah). This J source is the subject of Harold Bloom's inquiry.

The author that Bloom imagines for the Book of J is not a theologian or religious chronicler, but a talented and sophisticated dramatist who told stories purely for the joy of it. This writer, he suggests, was an aristocratic and well-connected woman living in the late Solomonic period (c. 950-900 B.C.), at a time when the United Monarchy was heading toward dissolution. Bloom further suggests that as later rationalist and revisionist strands meshed with J, they suppressed her lively representation of the ongoing interaction between God and humankind, in particular replacing her impish, irrational, and irascible God with a moral, perfect, and transcendent Deity. Therefore, in order to reconstruct J's unique perspective on the origins of Israel, it is necessary to pluck the J verses out of their Biblical context and read them on their own.

What leads Bloom to suppose a female author for the Book of J? His main argument is that J seems to have only heroines, no heroes. Sarah, Rachel, and Tamar (who, Bloom says, is the most memorable character in the Book of J) behave admirably, exhibiting clarity of vision and resoluteness of purpose, while Abram, Jacob, and Moses receive mixed reviews. As attractive as his full array of arguments is, I await further evidence. In Talmudic literature, presumably authored by men, a strong woman is often introduced to shame a weak man.

The centerpiece of Bloom's commentary, however, is not his theory of J's identity but an analysis of J's literary artistry. He singles out three features of her writing style for special praise: ellipsis, irony, and wordplay. Pointing again and again to instances of her ironic or comic stance, Bloom explains that the basis of J's irony is "the clash of incommensurates, a clash that begins in the illusion of commensurateness." Illustrations of this phenomenon are Abraham's wrangling with God about His contemplated destruction of Sodom and Gomorrah, and the people of Babel's striving to

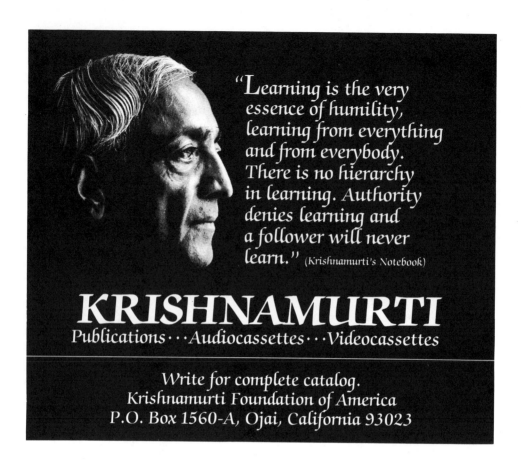
reach God in order to become men with a name.

What I find most appealing about this volume is Bloom's fresh reading of some major episodes of the Bible. His prose is so elegant, his literary perspective so broad, and his interpretive assertions so engaging, that it becomes almost irrelevant whether the hypothesis of J's authorship is probable, possible, or absurd. What flows from Bloom's unorthodox separation of J from the rest of the Bible is an opportunity to view the plots and characters (including God) in a new light. By articulating J's grand themes, such as homelessness and exile as a punishment for the overstepping of boundaries, he deepens the reader's appreciation of this set of stories, and even of biblical literature in general.

The middle section of *The Book of J* is a translation by David Rosen-

berg of all verses that scholars have attributed to J, but modified somewhat by Harold Bloom himself. This new translation, Bloom notes, attempts to preserve J's ironic stance and extended wordplay, without sacrificing the lyrical force of the Hebrew original. One device Rosenberg regularly uses is the substitution of "watch," "see," or "now look" for the omnipresent *"vayehi"* (and it came to pass). At the same time he changes most verbs from past to present, thereby drawing the reader into the narrative and conferring a sense of immediacy. Instead of the King James version's "And it came to pass, as they journeyed from the east, they found a plain in the land of Shinar; and they dwelt there" (Genesis 11:2), Rosenberg translates, "Watch: they journey from the east, arrive at a valley in the land of Sumer, settle there." What also characterizes this

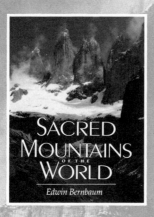
translation is its sparseness and simplicity, producing a text that reads and sounds more like the Hebrew original than the smooth translations of the last several centuries.

But what troubles me about this new translation are the frequent departures from literal meaning, which, in the absence of an accompanying explanation, seem to be mistakes. Why does Rosenberg translate "*vayishma Abram leqol Sarai*" (Genesis 16:2) as "Abram grasped Sarai's words"? It is true that in Biblical (as well as modern) Hebrew, "to hear" can also mean "to understand, to grasp," but when followed by "*leqol*," it can only mean "to obey, to heed," as in the Jewish Publication Society's 1982 translation, "Abram heeded Sarai's request." On what basis does he render "*bigdel Esau hahamudot*" (Genesis 27:15), as "Esau's clothes, those . . . *ready for washing*,"

when this phrase is usually translated as "Esau's *best* clothes"?

I am further baffled by Bloom's statement that Rosenberg preserves J's wordplays when time and again no such wordplay exists in J's Hebrew text. For instance, "bound," "boundary," and "unbound" in the Babel episode have Hebrew roots that are radically dissimilar: *nafuz* (unbound), *lo yibazer* (no boundary exists), and *vayahdelu* (came unbound). Even though what we see in these verses may well be variations on a theme, the absence of assonance suggests that one cannot reasonably call it a play on words. One begins to wonder to what extent Rosenberg's translation accommodated itself to Bloom's interpretation instead of giving rise to it.

The question Bloom seems to be asking throughout this book is whether the Bible is a great work only

because of the religious message breathed into it, in his opinion, by later revisionists, or whether, in its "unvarnished" form, it can still elicit kudos from secular, sophisticated individuals, like himself. His unequivocal answer, with which the reader is likely to agree, is that J's literary artistry surpasses even that of Shakespeare.

—*Judith Hauptman*

Judith Hauptman is associate professor of Talmud at the Jewish Theological Seminary of America and author of Development of the Talmud Sugya *(University Press of America).*

Sacred Mountains of the World
By Edwin Bernbaum. San Francisco: Sierra Club Books, 1990. Pp. xxiii + 291. $50.00, cloth.

A BOOK ABOUT such a universally appealing subject as mountains in their manifold religious and symbolic contexts, published by the same people who invented coffee table photo books on the environment, would seem a fairly predictable product. Accompanying scenic panoramas, we might expect large-type italicized evocative quotes extracted from Buddhist, Christian, or Islamic scriptures, and a workmanlike text serving as obligatory connective tissue for the picturesque lot—all delivered to your local bookstores in time for the holidays.

Happily and surprisingly, we would be quite mistaken. Ed Bernbaum is no armchair scholar and this book has been a profoundly personal adventure for him. It is far more ambitiously conceived, comprehensively researched, and skillfully designed than any holiday picture book. It reaches beneath the visible topography to examine the spiritual and philosophical traditions which underlie what we see, and to acknowledge the spiritual and moral authority of these mountains. In its sophisticated blend of text and visuals, the book is fully up to the task of introducing readers to the subsurface as well as the obvious attractions of our planet's grandest embodiments of the sacred.

A former Peace Corps volunteer, Bernbaum earned a Ph.D. in Asian Studies at the University of California in Berkeley and has written extensively about mythical and sacralized landscapes. He has also climbed many of these peaks; their awesomeness is ingrained in his muscle fiber. To his credit, however, Bernbaum wears neither his adventuring nor his academic credentials on his sleeve; one gets the sense that the book's ten-years-in-the-making represented his greatest mountaineering challenge. These mountains conquered him: the narratives he has synthesized, from research into their intimidating range of belief systems and ritual practices, never lose their flavor of wonder, modesty, and re-

spect. Bernbaum the mountain climber and Bernbaum the religious scholar join hands with Bernbaum the boy who adored high wild places and combine into Bernbaum our host on the serious, marvelous pilgrimage which this book is.

Instead of prefacing his work with an illustrated primer on the multifaceted roles of mountains in various religious traditions, Bernbaum wisely saves comparative discussions for his later chapters. First, he guides us in a series of giant steps across the sacred mountain ranges around the globe—in Tibet, China, Central Asia, Japan, Southeast Asia, the Middle East, Europe, Africa, North America, Latin America, Oceania. In each locale, we linger long enough to absorb exquisite details. We follow Milarepa in his mythic race against the Bon priest to the summit of Mount Kailas, holiest of Himalayan peaks to Hindus and Buddhists alike. We accompany pilgrims up T'ai Shan in China, rocky representative of the great god known as Emperor of Heaven, and we hear the Chagga of Tanzania tell how their inner traditions revolve around Mt. Kilimanjaro. Although dates, personages and references are abundant—and we have notes and a bibliography at the end—they never seem tedious cataloguing but are woven into rich description and good storytelling.

In two other ways the book manages to skirt becoming an illustrated encyclopedia. First, there is the comprehensiveness of its text. Whether we are exploring the monasteries and shrines of Israel's Mount Sinai or visiting the San Francisco Peaks with the Hopis and Navajos of the American

Southwest, we enjoy thick samplings of age-old belief systems. For each of the fifty or so specific peak or mountain range accounts, Bernbaum creates something between an "entry" and "essay," and efficiently leaves us slightly hungry for more as he moves us onwards around the world. Although the text is helpfully peppered with subheads, and we march through continent after continent, the presentation never seems obligatory or plodding. In the text's body, well-chosen, italicized quotes are printed small and are always too germane to Bernbaum's discussion to be merely ornamental or sentimentally poetic.

Secondly, the book is brilliantly illustrated and designed. Any initial dread at being overwhelmed by mural-like mountainscapes of places too grandiose for any but the most intrepid mountaineer is unfounded. The book respects yet controls its spirited peaks. There is a cunningly effective restraint behind the blend of micro close-ups and reduced-size macro mountainscapes. Mammoth promontories are mostly shown half page and do not "bleed" to the edge, hence never drowning the text. Human beings are continuously depicted, within or without mountain backdrops, so we never forget that what sacralizes these peaks are the cultural beliefs and personal feelings of human residence of these sacred geographies. A judicious use of mountain-related illustrations from folk art, tribal ritual, fine art, and odd historical sources keeps the visuals unpredictably intriguing.

In his final chapters Bernbaum ventures into the mountains of the soul, the mind, and the heart. He highlights themes from comparative religion in which mountains are seen as pivots of the world, as entry points into the divine, as corporeal bodies of great deities, as arenas for measuring spiritual growth, as analogies for ulti-

mate revelations. He opens up discussion of the universal impulse of poets, painters, photographers, and spiritual teachers to imagine and reproduce mountain metaphors for heaven and hell and struggle and hope. At the end, in strokes of unpretentious honesty, he returns to mountains as his first love. Without hectoring us into some physical or spiritual self-help campaign, Bernbaum makes his personal case for the transcendental dimensions of mountaineering and the inspirational importance of mountains in "everyday life." He handles his enthusiasms for high places with such disarming candor that our sense of an underlying contradiction between his post-traditional point of view that "The reality revealed by a mountain can be experienced anywhere. Mountains are no more sacred than any other feature of this landscape," and the highly specific beliefs of most of the cultural world views he has discussed, somehow does not matter. With this labor of devotion and apprenticeship to mountains and their diverse spiritual roles, he has earned the right to speak his mind.

—*Peter Nabokov*

Peter Nabokov is the co-author of the recent Native American Architecture *(Oxford University Press), and is writing a new book,* Sacred Geography: Patterns and Passions on Earth *(Harper & Row).*

Freedom in Exile: The Autobiography of the Dalai Lama
New York: HarperCollinsPublishers, 1990. Pp. *xiv* + 288, $22.95, cloth.

IN 1962, JUST three years after his flight from Tibet, the twenty-seven-year-old Dalai Lama published his autobiography, *My Land and My People*. Originally written and released in Tibetan and later translated

The Hunter
and the Hunted

A New Look at
an Ancient Metaphor

Presented by

THE SOCIETY FOR THE STUDY

OF MYTH & TRADITION

in association with

THE AMERICAN MUSEUM

OF NATURAL HISTORY

Wednesday, May 8, 1991 at 7 P.M.

in the Main Auditorium of the Museum

Central Park West at 79th Street, New York City

The speakers will include:

LINDA HOGAN, Chickasaw, author of *Mean Spirit* (Atheneum)

RICHARD NELSON, author of *The Island Within* (Vintage Books)

and other participants to be announced.

The symposium will coincide with PARABOLA's Summer 1991 issue, which explores the theme of THE HUNTER in various traditions as a symbol of risk, survival, and spiritual search.

Tickets are free to PARABOLA subscribers and members of the Museum. Up to two tickets can be requested by each subscriber by writing to the magazine and including the subscription number and magazine label from the current issue, along with a stamped, self-addressed envelope, to: PARABOLA, 656 Broadway, New York, NY 10012.

A limited number of tickets may be available to non-members for $4.00 each. For ticket availability, call the Museum office at (212) 769-5606 after April 15.

This event is being held in cooperation with Vintage Books.

into English, the work recounted his extraordinary childhood, including his being discovered at the age of three to be the reincarnation of the thirteenth Dalai Lama, the next-in-line in the series of human embodiments of Avalokiteśvara, the bodhisattva of compassion, who have been Tibet's rulers since the seventeenth century. He told of his investiture, his monastic education, and of the dark events that led to the occupation and annexation of Tibet by the People's Republic of China. Since the publication of *My Land and My People*, many books have been written about Tibet and the Dalai Lama, the least interesting recounting the same material with little additional information and much less insight; the best, like John Avedon's *In Exile from the Land of Snow*, providing significant background to our picture of this most fascinating man. And with his being awarded the Nobel Peace Prize in 1989, the books continue to appear.

It is therefore a great pleasure to report that the Dalai Lama, now fifty-five, has composed a new autobiography, and it is by far the best book about him so far. Much of the same material is covered as in the previous autobiography, but with much greater detail and candor. For example, he says that he remembers very little of the long and elaborate ceremony in which temporal control of the nation

was passed into his hands, except the discomfort of a full bladder. But in other areas he displays a remarkable memory, recounting in detail his meetings with such figures as Mao Tse-tung, Chou En-lai and Nehru. The book is rich in historical detail concerning the fall of Tibet in the 1950s, with the Dalai Lama reporting events and conversations heretofore unknown. The work will therefore be of signal value for historians of Asia. At the same time, this is the autobiography of one of the truly distinguished Buddhist scholars of this century, and the descriptions of his education under the tutelage of the savants of Tibetan Buddhism provide fascinating insights for the many western followers of that tradition.

Perhaps above all, the autobiography tells the story of one of the most remarkable lives of our century. Here is the captivating tale of a peasant boy from a culture popularly conceived as one of the most obscure in the world, elevated to a position invested with the highest worldly and spiritual powers. Our expectations are both confirmed and confounded, often in delightful ways: the Dalai Lama describes his chambers in the Potala not as sumptuous and regal but as dark and dank, where offerings to the buddhas and bodhisattvas arrayed on his altar quickly become offerings to the hordes of mice that populate his palace. He writes affectionately of his family, especially his brother Lopsang Samten, with whom he seems regularly to have engaged in fisticuffs. It was Lopsang Samten who held the dubious distinction of being the Dalai Lama's sole classmate:

On opposite walls hung two whips, a yellow silk one and a leather one. The former, we were told, was reserved for the Dalai Lama and the latter was for the Dalai Lama's brother. These instruments of torture terrified us both. It took only a glance from our teacher at one or other of these whips to make me shiver with fear. Happily, the yellow one was never used, although the leather one came off the wall once or twice. Poor Lopsang Samten! Unluckily for him, he was not such a good student as I was. But then again, I have a suspicion that his beatings must have followed the old Tibetan proverb: "Hit the goat to scare the sheep."

The autobiography takes us right up to the present, describing the Dalai Lama's recent trips to the West, including his triumphant visit to Czechoslovakia, where Vaclav Havel was the first head of state to offer him an official state invitation. Over dinner, the unaffected Havel, with a beer in one hand and a cigarette in the other, told his guest that he felt a special affinity with the sixth Dalai Lama (who had a reputation of being something of a libertine). The Dalai Lama also describes in some detail the current situation in Tibet and his indefatigable efforts to restore human rights to the Tibetan people.

As has become clear from his numerous publications over the past decade, the Dalai Lama is also one of the most creative thinkers to appear in the Tibetan Buddhist tradition, due at least in some measure to his forced confrontation with the modern world. Thus, his reflections on reincarnation and on the efficacy of Tibetan medicine, as well as his description of his personal relationship with the deity

that periodically inhabits the body of the state oracle of Tibet, are particularly provocative.

The book, then, is of value to students of international relations, of Tibetan cultural history, and of Tibetan Buddhism and to everyone who is fascinated by this extraordinary man who rises at four and spends the hours until noon in meditation and study, who has a passion for gardening and is not reluctant to use his air rifle to keep crows away from his bird-feeder, all while working in ways seen and unseen for human liberation.

—*Donald S. Lopez, Jr.*

Donald S. Lopez, Jr. is professor of Buddhist and Tibetan Studies at the University of Michigan.

The School of Charity: The Letters of Thomas Merton on Religious Renewal and Spiritual Direction

Brother Patrick Hart, ed. New York: Farrar, Straus & Giroux, 1990. Pp. *xiii* + 422. $24.95, cloth.

N AMED FOR A term taken from St. Bernard of Clairvaux referring to the monastery, *The School of Charity* is the third of a projected five-volume series of letters by the celebrated Trappist monk, Thomas Merton. Although these letters are written by a man deeply involved in the life of his monastery of Gethsemani in Kentucky, they are also profoundly concerned with renewal on a larger scale, as the subtitle connotes: an interior regeneration in the Catholic Church, in religious communities, and in society as a whole.

Patrick Hart, who was Merton's secretary, has done a superb job of editing this volume of fascinating letters (each volume has a separate editor and draws from the full twenty-seven years of Merton's monastic experience at Gethsemani, from 1941 to 1968). Merton's time as a monk was exactly half of his total life span, and it was during this period that he matured as a writer, spiritual thinker, cultural critic, and moral voice, one venerated by millions around the planet. His literary productions took form in some forty books, two thousand articles, and hundreds of book reviews. But a writer's personal letters are also a significant part of his or her literary output and a measure of the inner landscape.

The letters in this volume are arranged into three chronological divisions: (1) "The Early Monastic Years" (1941-1959), (2) "The Middle Formative Years" (1960-1964), and (3) "The Later Solitary Years" (1965-1968). Jean Leclercq, a French Benedictine monk and writer; Dom James Fox, Merton's abbot; and Dom Gabriel Sortais, the Abbot General of the Trappists, were among his longstanding correspondents. The latter two served as spiritual fathers to whom Merton confided the secret workings of his soul and the struggles of the interior life, particularly concerning the issue of solitude. In Jean Leclercq he found a dear and faithful friend and a fellow scholar and writer with whom he could share his intellectual concerns as well as his deep longing for solitude.

This love of solitude runs as a constant theme throughout these letters and reveals the simplicity and depth of Merton's spirit. He hated publicity, his celebrity status, the endless details associated with publishing, and the correspondence attending his rise to the public eye. He was wonderfully affable and even outgoing, so much so that one person once remarked that "Merton would love to be a hermit, but in Times Square"—a remark that seems unfair and inaccurate.

Merton's desire and need for solitude were too deep, indeed constitutional to his nature and spiritual life. In a letter to Leclercq in May 1959, he confided: "I naturally keep a certain desire for solitude in my heart and cannot help but hope that some day it may be realized." And in another letter to Leclercq in November of the same year, he brings up the question of solitude within the context of a possible Cistercian or Trappist foundation in Latin America that he dreamed would include himself. Here he reveals the nature of his desire for a solitary life:

. . . I still have hope that at the same time they [his superiors] will let me go to make the experiment. . . . I am who I am and always have the writer's temperament, but I am not going down there to write, nor to make myself known, but . . . to disappear, to find solitude, obscurity, poverty. To withdraw *above all* from the collective falsity and injustice of the U.S. which implicate so much of the Church of this country and our monastery.

Merton's extraordinarily far-ranging interests in politics, peace, justice, race relations, the nuclear issue, the anti-war movement, monastic renewal, spiritual direction, liturgical reform, and social evolution, as well as his espousal of Pasternak and other martyrs of freedom, all hide the essential simplicity of his character. He was complex in his witness as writer, poet, literary critic, and prophet, but he was also a simple monk fighting the same inner battle with egotism as his fellow monastics. With it all, he was humble, obedient, even docile, though he loathed censorship. At least this is the portrait his letters paint; they uncover this quality of simplicity and honesty as stable elements in his seemingly complex character.

John Eudes Bamberger, the abbot of the Genesee Abbey in Piffard, New York, who had been a student under Merton at Gethsemani, made a telling observation about him five years ago

in an essay-review of Michael Mott's biography, *The Seven Mountains of Thomas Merton*, in *Cistercian Studies*. Abbot Bamberger lamented what he regarded as a misunderstanding of Merton that arose by emphasizing one aspect of his life over another. For him, Merton was always a mystic, and the mystical dimension is the key to appreciating this sometimes bewildering figure. It also explains Merton's profound interest in the East and its mystical teachings. His own journey to Asia at the end of his life was actually just another chapter in his inner drama as he approached the Divine.

The School of Charity inspires because it puts us in contact with a figure who symbolizes our century's complexity and its longing for transcendence. Thomas Merton, in his work and life, points both to the transient nature of this complexity, and to the reality of something permanent that can only be discovered when the spiritual sojourner finds his or her home in the eternal.

—Wayne Teasdale

Wayne Teasdale, a monk and writer, is the author of Essays in Mysticism *(Liturgical Publications), and* Towards a Christian Vedanta *(Asian Trading).*

What is Civilisation? and Other Essays

By Ananda K. Coomaraswamy. Great Barrington, Massachusetts: Lindisfarne Press, 1989. Pp. *xiv* + 193. $14.50, paper.

IN THE GENERATION that came of age in about 1900, a number of scholars and seekers, Ananada Coomaraswamy (1877-1947) among them, glimpsed the same historic task. The challenge, in essence, was to renew Western religious and philosophical life by a transfer of knowledge from Asia to the West. The moment was propitious: the traditional cultures of Asia remained what they always had been in many regions, although colonialism and industrialism were taking their toll, while Western culture was divided between complacency on the one hand and fierce self-questioning on the other.

Born in Ceylon, the son of a distinguished Tamil legislator and his English wife, Coomaraswamy was educated in England, where he took a degree in geology at London University and then went back to Ceylon as director of a mineralogical survey. Traveling that Eden in search of mineral deposits, he found also the deposits of its ancient Buddhist culture and discovered his vocation as a dedicated expositor of traditional cultures: their arts, religions, philosophies, and social forms, their unison on first principles beneath the luxuriant diversity of manifestations, and not least their vulnerability to the incursions of the secular, industrial cultures of the West. His first major book, *Mediaeval Sinhalese Art* (1908), initiated themes that would only deepen over the next forty years of his extraordinarily prolific career as a museum curator and author.

In the years before World War I, Coomaraswamy built an unexcelled collection of North Indian painting and moved to Boston in 1917 to found in that city's Museum of Fine Arts the first department of Indian art in any American institution. In the year following his emigration, he published in the United States a book of Indian essays, *The Dance of Shiva*, which to this day remains in print and seems the freshest word on its topics.

Yet all of this was early work. The man changed profoundly in the late 1920s and went on to create a canon of works which continued the

logic of his prior efforts in a wholly unpredictable, and unpredictably grand, manner. In the present collection, *What is Civilisation? and Other Essays*, the chapter entitled "On Being in One's Right Mind" suggests the nature of that change:

Metanoia, usually rendered by "repentance," is literally "change of mind" or intellectual metamorphosis. Plato does not use the word, . . . but certainly knows the thing: for example, in *Republic* 514F, the values of those who have seen the light are completely transformed. . . . The notion of a change of mind presupposes that there are two in us: two natures, the one humanly opinionated and the other divinely scientific; to be distinguished either as individual from universal mind, or as sensibility from mind, and as non-mind from mind or as mind from "madness"; the former terms corresponding to the empirical Ego, and the latter to our real Self, the object of the injunction "Know thyself."

In his later years Coomaraswamy created in essay after essay, book after book, a synthesis that drew equally from ancient Greek and Sanskrit sources, from the Latin of St. Thomas Aquinas and the nascent German of Meister Eckhart, from Rumi's Persian, and all languages of the world that spoke through the centuries, wisely, of Truth and "the pilgrim's way" (the title of an essay in the present collection). Parallel to this exploration of traditional religious and philosophical literature, he explored no less thoroughly the nature of traditional religious art, and the links between art and knowledge, in all of the developed cultures of the world—among which he counted numerous minority cultures, including that of Native Americans.

What Is Civilisation? and Other Essays is assembled from the works of these later years. A permanent contribution to the Coomaraswamy bibliography and carefully prepared in all respects, this book returns to print essays such as "The Symbolism of Archery" and the pair on "Gradation and Evolution" which are among his finest. In addition it gives first publication to a number of essays written toward the close of his life. Virtually the entire *theory* of the Way can be found in his writings, demonstrated with magisterial finality, and to the extent that lucid thinking already represents a step forward on the Way, there is more here than theory. In Coomaraswamy's recension, the Way to liberation is daunting, its intellectual and moral demands immense; yet behind his erudite words, one hears the invitation to which he himself was responding:

What is then from the standpoint of metaphysics the whole course of an individual potentiality, from the "time" that it first awakens in the primordial ocean of universal possibility until the "time" that it reaches the last harbour? It is a return into the source and well-spring of life, from which life originates, and thus a passage from one "drowning" to another; but with a distinction, valid from the standpoint of the individual in himself so long as he is a Wayfarer and not a Comprehender, for, seen as a process, it is a passage from a merely possible perfection through actual imperfection to an actual perfection, from potentiality to act, from slumber (*abodhya*) to a full awakening (*sambodhi*).

For some in the realm of actual imperfection, these dry but expansive words, so characteristic of their author, have the repeated awakening snap of a prayer flag aloft in the wind.

The migration of knowledge from Asia to the West and its assimilation by thoroughgoing comparison with our own traditions were the substance of Coomaraswamy's lifework. The historic cycle to which he contributed concluded tragically, but not without positive consequences, when Tibetan religious culture was exiled from its homeland, but found new

homes not only in north India but in Europe and America.

— *Roger Lipsey*

Roger Lipsey, a contributing editor to PARABOLA, *is the author of* Coomaraswamy: His Life and Work *(Princeton University Press) and* An Art of Our Own *(Shambhala).*

Mystical Islam: An Introduction to Sufism

By Julian Baldick. New York: New York University Press, 1989. Pp. 208, $40.00, cloth; $15.00, paper.

THE OPENING discussion of the origins of Sufism in *Mystical Islam* is a notably broad-based treatment of the topic for an introductory text. Highly speculative possible influences are weighed and discussed, even if briefly, from Judaism, neo-Platonism, shamanism, Manichaeism, Mazdaism, and Indian religions, as well as testimony from the Messalians, Ebionites, Cynics, and Gnostics. This breadth is matched in both the temporal and geographical scope of the work, which laudably includes an examination of Sufism in the twentieth century and in areas not usually covered in introductory surveys (Central Asia, Indonesia, West Africa). In addition, Julian Baldick provides his reader with some sense of the controversies current in western academic literature on Sufism, drawing into his own discussion other researchers, their works, and some of their ongoing disagreements, though not always in an even-handed manner. Footnotes have been kept to an absolute minimum, and a useful bibliography of further readings steers the interested student toward classic studies on a variety of topics.

The bulk of this work, however, is not so much a general introduction to Sufism as it is a chronological history of Sufi writers and thinkers. In the three central chapters of the book, over eighty of these are dealt with, one after the other, in chronological order, in the space of 130 pages. Most of them are dispensed with in no more than two or three paragraphs. As the first paragraph in the treatment of each figure is primarily biographical, the remaining summaries of their thought must necessarily be highly condensed, painstakingly selective, and sometimes superficial. Little or no synthesis, in the form of discussion of major themes or developments, is presented to organize the resulting kaleidoscopic effect or to ease the reader's path. In only a few cases are specific writings described or examined; though Sufi "manuals" are mentioned at many points, for example, the reader is never told what such a manual actually comprises.

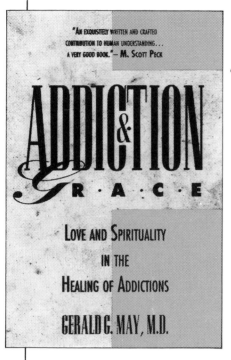
This book is purposefully devoted to discovering "concrete instances of the repetition of patterns, expressed explicitly in literary texts or material remains, with such clarity of definition as to make acceptance of historical continuity beyond dispute." The strictly linear, chronological presentation of the writings of Sufi thinkers functions, however, in an unfortunate manner to undermine Baldick's general aim: instead of foregrounding and highlighting continuities or contradictions, it buries them as scattered details in the "who's who" format of the book. Themes and patterns surface and disappear in a seemingly random fashion due to the brevity of the discussion of each individual. Key concepts at times appear suddenly at strange points in the chronology when they have in fact been present in the tradition for several centuries. For example, there is an abrupt mention of the *qutb* "pole" or "chief mystic" in the discussion of the fourteenth-century writer Ibn Khaldûn, as well as a brief reference to initiation by bestowal of the *khirqa* "patched frock" in the discussion of thirteenth-century brotherhoods. No uninitiated reader will be able to deduce that these are significant, even basic, images in the vocabulary of Islamic mysticism.

The author's constant concern to delineate influences and precursors often overwhelms the introductory nature of the book. The four paragraphs devoted to 'Alâ' al-Dawla Simnânî are typical: the first paragraph presents biographical data, the next examines possible Shiite influences, the third possible Buddhist and Taoist influences, and the final paragraph claims discernible Indian influences. The only detail of Simnânî's thought actually presented in the passage is that he was "particularly preoccupied by visions of coloured lights, and he is the most important systematizer of this aspect of Sufi experience."

Remarkable precursors often emerge in the discussion haphazardly and to no clear purpose. For example, the "three-fold tying of the belt" in a Turkish Bektashi initiation ". . . corresponds to the Indo-European triad of religious knowledge, force and fertility. . . ." The question of the significance of such continuities, even if they were to be accepted *prima facie*, is consistently left begging.

After tracing nearly all the elements of Sufism to earlier sources, the reader might well be forgiven for wondering if Islamic mysticism is in fact nothing but an unoriginal amalgam of ingredients from earlier, and presumably more creative, cultures, particularly when the author virtually states as much in his conclusion. Posing his final question about whether Islam could exist without Sufism, the author decides:

> The historical evidence suggests that it could not. For the religious core of Islam seems, from very early on, to have been a blend of Jewish law and Christian devotionalism, with a Gnostic element already present. If the Christian part [*i.e.*, Sufism] were to be suppressed what would remain would be a Judaism without Jews, a national religion without ethnic identity.

Though the author has clearly targeted as readers "those who have read nothing about Islam or mysticism," *Mystical Islam* will probably only function effectively as an introduction in conjunction with a more thematically oriented text or, alternatively, as a historical review, with a rigorously narrow focus and an impressive historical scope, for readers who already possess some familiarity with the field.

—*Dwight F. Reynolds*

Dwight F. Reynolds is professor of Arabic language and literature at Amherst College.

The Yoga of the Christ in the Gospel According to Saint John
By Ravi Ravindra. Shaftesbury, Dorset: Element Books, 1990. Pp. 234. $15.95, paper.

CHRIST'S YOGA? The very idea seems bizarre, if we think of yoga as merely physical exercises to stretch the body. But if we see yoga as a mental discipline as well and give it its true meaning of union—in this case, union with the consciousness of the Christ, with the I AM state in

which "I and the Father are one" — then we are already in the spirit of that greatest and most elusive sacred text of Christianity, St. John's Gospel.

A physicist and metaphysician, Ravi Ravindra is uniquely qualified for the task of exploring the "yoga of the Christ." Brought up in India, he moved in 1961 to Canada for postgraduate work; currently, he is a professor of both physics and comparative religion at Dalhousie University in Halifax, Nova Scotia. As an Indian, Ravindra looks at St. John's Gospel as an outsider does, in much the same way that a native Canadian, for example, might look at the great traditional texts of India. One can always uncover fresh understandings of one's own tradition by looking at familiar sacred texts in the light of another ray from the same Sun.

Ravindra's study (which readers of PARABOLA have already had an opportunity to sample in Vol. XIII, No. 2) shows once again the benefits that cross-cultural perspectives can bring. By immersing themselves in another language, another culture, another tradition, people can suddenly become aware that it's possible to see themselves and the world differently: they now have stereoscopic vision, whereas before they saw with only one eye. Ravindra helps the reader to open both eyes, providing one sees—as the author constantly urges—with the heart as well as the head.

Ravindra has managed to share his insights into the essential meaning of St. John's Gospel without being diverted by sectarian or theological controversies. He has made his own selection, adapted from the many translations available in modern English idiom, and he takes us through the entire text, section by section, with an extremely helpful commentary. On many occasions, he is illuminating— for example, in his discussion of the roles of Judas, Lazarus, and Mary Magdalene (although I have difficulty agreeing with him that "the disciple whom Jesus loved" was Lazarus, when it is stated that this disciple is the author of the Gospel). But overall, this is the best study of St. John's Gospel that I have found, though one does wish that Ravindra repeated himself less.

To convey something of the flavor of the work, here is Ravindra's description of Mary Magdalene's being the first person to see the risen Christ:

She was there when he was crucified, and she is there now when he rises from the dead. He has undergone enough transformation that she has difficulty recognizing him, until he calls her by name. In her response, there is not the least hint of fear or hesitation; there is only pure joy, and she bursts out with a term of endearment, *Rabunni*—my dear Rabbi! The translation of the Aramaic word *Rabunni* as "Master" ignores the feeling of tenderness associated with this diminutive of Rabbi. . . .

It is difficult to read this chapter in the Gospel without wondering if Mary was the feminine balance to the masculinity of Jesus, the yin to his yang, as Radha to Krishna.

By respecting not only the parallels but also the differences between traditions, Ravindra avoids the error of equating terms used in different traditions for higher states: states which, just because they are higher, may not neatly fit our lower-level definitions. Perhaps the greatest difficulty in a study of this kind is the temptation to share one's insights in too literal a way—"this" means "that." If understanding is reached only with the head, the ultimate mysteries are inevitably reduced. But if the heart is open and the feeling is touched by something higher, what seems obscure to the mind may reveal itself in a new perception of meaning.

As Ravindra writes in his introduction, "Anyone who speaks from that core of oneself, which is possible only when one has surrendered all of one's relatively superficial selves to the service of this one Self, constitutes a bridge, a way, from here to There." For the reader ready for such a non-judgmental journey into the interior, *The Yoga of the Christ* will stretch the understanding of oneself, as well as of St. John's Gospel.

—James George

James George served as Canadian ambassador in India, Nepal, and Sri Lanka for nearly ten years and was the founding president of the Threshold Foundation in London.

Words With Power. Being a Second Study of "The Bible and Literature."

By Northrop Frye. San Diego and New York: Harcourt Brace Jovanovich, 1990. Pp. *xii* + 274. $24.95, cloth.

W E ARE suffering the results of a century-old rejection of the Bible as the source of anything illuminating. But however remote the Bible may seem to us today, Northrop Frye shows us that our society, like all others, has a mythology which is woven so subtly into the fabric of our culture that its threads

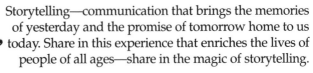
are taken for granted. More than this, as he demonstrates in *Words With Power*, our language itself, even beyond its Biblically informed imagery, has been shaped by the thought-processes that the Bible has given us. The impact of the Bible on our mythology and on the semantic subtext of our thinking is seldom acknowledged, concerned as we are with repudiation of the text; Frye's work is that acknowledgement.

Frye has devoted considerable effort to the study of the impact of the Bible, not as literature, but on literature. It is a theme which he initially addressed in his *Anatomy of Criticism* in 1957, and developed more fully in *The Great Code* (1982), which also bore the subtitle "The Bible and Literature." In the latter work he explained that he was approaching the Bible as a literary critic, and that his study was neither a work of Biblical scholarship

nor of theology. "What matters," he wrote in the introduction to *The Great Code*, "is that 'The Bible' has traditionally been read as a unity, and has influenced Western imagination as a unity." The Bible has a fundamentally historical structure, starting with creation of the universe and ending with a vision of its end. In between lies the history of the human spirit, elaborated by means of concrete and essentially experiential imagery as well as by an extraordinary symbolism. It was this aspect of the question that Frye addressed in *The Great Code*.

In the present study, Frye first questions the distinctive social function of literature and the basis of the authoritative voice of the poet. He then examines one particular image which has deeply affected our literature and our thinking, that of the *axis mundi*, or world axis. This is an especially significant symbol because it is essentially a verbal construct and because it occurs in other contexts than the Biblical, though it is the Biblical version which has had the greatest impact on us. It is the source not only of the dualism by which we categorize things as "good" and "bad," and distinguish between subject and object, but also of our notions of hierarchy. The willing spirit and the weak flesh are much more intricately related than a simplistic contrast would at first suggest.

One question which this book raises is how much of the impact on literature is made by the Bible itself, or how much comes from the King James version: how much of our cosmological imagery is in correspondence with the centuries of religious tradition preceding the Reformation, and how much with that of the sixteenth-century humanistic revolt. The Elizabethan view of the cosmos derives from a number of sources, not all of them Biblical, due to the humanistic rejection of the intellectual limits imposed by the Church. It was in this milieu that Shakespeare wrote, and in which the King James translation was made. Perhaps we do not think the Bible itself was written in the language of Shakespeare, but that is the form with which we are most familiar, and there is a tendency in our thinking to prefer that version to any other.

Frye's examination of the different modes of linguistic expression employed in the Bible and in our literature seems to show how those writers, chiefly poets, whom we most venerate are those who made explicit use of Biblical imagery and expression; also, how the ordering of what Frye calls "the imaginative cosmos" underlies virtually all our judgments of value, judgments which carry considerably more weight and have much more effect on our behavior than we think.

Each successive generation since the advent of the scientific revolt of the nineteenth century has abandoned the Bible more and more: first as the source of all knowledge, metaphysical as well as technical, of the cosmos and man's place in it, latterly as the source of a teaching that could inform our lives and at least sensitize us somewhat to our moral failings. What Frye shows us is that the Bible has influenced our thinking, and continues to do so, in ways more subtle than our ability to cite it or our acceptance of it as a guide to religious life.

—Paul Jordan-Smith

Paul Jordan-Smith is a freelance writer and senior editor of PARABOLA.

The Feathered Sun: Plains Indians in Art and Philosophy

By Frithjof Schuon. Introduction by Thomas Yellowtail. Bloomington, Indiana: World Wisdom Books, 1990. Pp. *xx* + 170. $37.50, cloth; $25.00, paper.

JUST WHAT IS the "feathered sun," that mysterious symbol that radiates in red and blue shafts on the dust jacket and also stands embossed on the sand-tinted cover of this hefty (11 ¼″ x 9″, 3 lb.), beautifully-bound volume? According to the author, it is a representation of "the Great Spirit in general and the Divine Presence in particular"—and, as such, a fitting symbol as well to find stamped throughout the pages of the latest book in Frithjof Schuon's (b. 1907) long, distinguished career as an advocate of the primacy of symbol, the urgency of esoteric understanding, and the ever-presence of divinity.

In many ways, this volume seems a culmination of Schuon's life-work. It gathers together on the theme of Native American religion nine of his most profound (and most difficult) essays, along with extensive excerpts from his correspondence and diaries. It also displays, to my knowledge for the first time between cloth covers, nineteen reproductions of his small, iconographic oil paintings on Native American themes.

Of the essays collected here, all originally written in French, six have previously appeared in English translation in journals or books. The first in the text, "The Symbolist Mind," is exemplary, laying out Schuon's core belief that symbols "at the origin—the 'Divine Age'—were in themselves fully adequate to transmit metaphysical truths" without the verbal explications we moderns find so necessary. This has particular application to Native American thought, for Schuon sees what he likes to call "Red Indians" as standard-bearers of "the symbolist vision of the cosmos." This vision, he goes on to explain, is a "spontaneous perspective that bases itself on the essential nature—or the metaphysical transparency—of phenomena." Thus, for example, for the Lakota Sioux, water is "the sensible appearance of a principle-reality" or *wakan*.

This same essay also advances one of Schuon's key points as a scholar of comparative religion: that the Plains Indians are not polytheists, but rather believers in a "Supreme God," which he identifies with the "Great Spirit" or (in Lakota) *Wakan Tanka*. This view accords nicely with Schuon's school of intellectual Sufism, which imagines the One Supreme Reality as manifesting Itself in different ways in different cultures. It should be noted, however, that Schuon's description of Native Americans as "primordial monotheists" remains controversial, not universally agreed upon by ethnographers or by Native Americans themselves, and that the author's forceful arguments in its favor are somewhat undercut by his tendency, which

Storytime Series

THE BOY WHO LIVED WITH THE BEARS AND OTHER IROQUOIS STORIES

Told by Joe Bruchac

SIX NATIVE AMERICAN TALES

Rabbit and Fox

The Boy Who Lived with the Bears

How the Birds Got Their Feathers

Turtle Makes War on Man

Chipmunk and Bear

Rabbit's Snow Dance

This collection of six Native American tales is the first of a new series which offers children of all ages traditional stories told by traditional storytellers.

The myths, legends, and folktales passed on through storytelling hold valuable messages from our ancestors, messages that are easily lost in the babble of the modern world. The *Storytime* Series was created to give back to children their stories, and with the promise that the stories offered are good and true.

These six stories, from the Iroquois people, or "Haudenosaunee," which means "People of the Longhouse," are of the adventures and misadventures of rabbit and buzzard, turtle and bear, and of one small boy who is befriended by all the animals of the forest. Between the stories the listener will hear drumming and the spirited "call and response" between the storyteller and his audience.

Joe Bruchac—or *Sozap*, as he is known under his Abenaki name—is a well-known storyteller and writer of Abenaki, English and Slovak ancestry who lives in the Adirondack foothills of New York where he learned many of the stories he tells on this tape.

Total running time: 63 min. $9.95
ISBN 0-930407-19-9

Please use the order form bound between pages 96 and 97.

Society for the Study of Myth and Tradition

Publisher of PARABOLA Magazine and PARABOLA Books

The Society for the Study of Myth and Tradition is a not-for-profit organization devoted to the dissemination and exploration of materials relating to the myth, symbol, ritual, and art of the world's religious and cultural traditions.

In addition to publishing PARABOLA Magazine and PARABOLA Books, the Society sponsors PARABOLA Events, a series of lectures, programs, and symposia. The Society also recently has begun offering video and audiotapes related to its field of interest.

The Society gratefully acknowledges the following contributors whose support has helped make its activities possible:

> Consolidated Edison of New York
> New York Council for the Humanities
> New York State Council on the Arts
> The American Museum of Natural History
> The Marsden Foundation
> The New York Community Trust

Contributions to the Society are tax deductible. Individual contributions should be directed to: Society for the Study of Myth and Tradition, 656 Broadway, New York, NY 10012.

Board of Advisors

Joseph Epes Brown
Writer

Jean Erdman
Co-founder with Joseph Campbell, The Theater of The Open Eye

John Fowles
Writer

Frederick Franck
Artist and Writer

Marie-Louise von Franz
Jungian Analyst and Writer

Stella Kramrisch
Curator of Indian Art Philadelphia Museum of Art

Ursula K. Le Guin
Writer

Lobsang Lhalungpa
Writer, Translator of Tibetan Texts

Helen M. Luke
Director Apple Farm Community

The Very Rev. James Parks Morton
Dean The Cathedral of St. John the Divine

Joseph Needham
Director, Science and Civilization in China Project, Cambridge

Richard R. Niebuhr
Hollis Professor of Divinity Harvard University

Dorothy Norman
Writer

Isaac Bashevis Singer
Writer

Huston Smith
Professor Emeritus Philosophy & Religion Syracuse University

Barre Toelken
Folklorist

P.L. Travers
Writer

David Wilk
President, Inland Book Company

runs entirely away from him now and then, to equate all Native American beliefs, and even all world-wide tribal beliefs, with those of the Plains Indians.

The other essays elaborate the basic theme of the primacy of symbol. "A Metaphysic of Virgin Nature" examines the recurrent Native American image of a cross inscribed in a circle with a man in the center, seeing this human figure as "mediator between the Earth and the Sky." This essay also affords some of the surprising insights that pepper this book and that can startle the reader into new understandings, such as Schuon's analogy between totemic animals and guardian angels, or the links he uncovers between Native American beliefs and those of the Bhagavad Gita, with their shared emphasis on the marriage of combat and contemplation.

The second essay, "The Shamanism of the Red Indians" draws further parallels between Native American monotheism and the Vedic idea of Brahma, and also serves as a primer for Schuon's ideas on the methodology of shamanism. "The Sacred Pipe" looks at the calumet as "a symbol and 'means of grace' of the first import," while "Symbolism of a Vestimentary Art" scrutinizes Native American dress, both in its imagery (especially the ubiquitous rosette with porcupine quills, a representation of the sun) and its cut (the equally ubiquitous fringe on jackets and skirts, which for Schuon symbolizes "the spiritual fluid of the human person"). As for the other essays, "The Demiurge in North American Mythology" examines the *heyoka* phenomenon, the subject of "The Sun Dance" is self-explanatory, "A Message on Indian Religions" argues that Native Americans understand nature as a manifestation of God, and "His Holiness and the Medicine Man" records a strange encounter, with Schuon as intermediary, between a Native American and a Hindu sage.

After these dense, hyperintellectual explorations, Schuon's diary entries and correspondence come as a relief. We hear the man talking in earthier tones, recalling visits to the Sioux, Crow, Cheyenne, and Blackfeet peoples, and attendance at two Sun Dances. The correspondence in particular strikes a wistful, sometimes deeply moving note, as when Schuon remarks that

. . . it is difficult to bear the sight of all the injustices in the world. But this earthly world will disappear anyway, and the most important thing is always spiritual life; if we cannot change the world around us, we can at least change ourselves. . . . I just want to point out that there is more reality in the Unseen above than on this visible earth, and that God alone is absolute Reality.

Alas, Schuon's brilliance as a metaphysician is but imperfectly reflected in his paintings. The foreword points out that these oils follow "traditional rules," such as an avoidance of foreshortening and shading and a loose hand with perspective. Be this as it may, most of the canvases, while valiantly trying to emulate the pure geometry of Native American designs, seem too schematic, too stiff to qualify as great art. Almost all feature dusty reds and dusky yellows, with heavily-blocked Indian faces reminiscent in some ways of Roualt. They seem like cousins to Native American art, cousins in which the blood-line has somehow thinned—although a few canvases, most notably *Cavalcade* (1961), achieve notable movement and majesty. Perhaps expecting great visual art from a great theoretician is one expectation too many: let us rest content with Schuon's words, which may help point the way to, as he says, "union with the Everlasting."

—*Philip Zaleski*

Philip Zaleski is a freelance writer and senior editor of PARABOLA.

CREDITS

Cover "The Moneychanger and his Wife" by Quinten Metsys, Flemish painter, 1465-1530. Musée du Louvre, Paris. Photograph © Scala/Art Resource, N.Y.

Page 5 Alice Spott, Yurok Indian, in ceremonial dress, about 1910. Photograph: Humboldt State University Library.

Page 7 Robert Spott, Yurok Indian, ca. 1920. Photograph by Coleman Studio, Oakland, Cal. Collection Thomas Buckley, Boston.

Page 9 From A.L. Kroeber, *Handbook of the Indians of California* (New York: Dover Publications, Inc., 1976).

Page 10 The *śakti* triangles. From Heinrich Zimmer, *Artistic Form and Yoga in the Indian Cultic Image* (Berlin, 1926).

Page 12 Illustration © 1991 Tom White.

Page 16 Borobudur, Java, 8th-9th century A.D. From Joseph Campbell, *The Mythic Image* (Princeton, N.J.: Princeton University Press, 1974).

Page 20 Amsterdam, 17th century. From The Jewish Museum, London. Photograph: Roland Sheridan. From Geoffrey Wigoder, ed., *The Encyclopedia of Judaism* (New York: Macmillan Publishing Company, 1989).

Page 23 Photograph by Mary Ellen Mark. Courtesy of Petrie Productions, New York.

Page 25 1869 engraving after a drawing by Theodor Pixis. Bildarchiv Preussischer Kulturbesitz, Berlin. From Charles Osborne, *The World Theatre of Wagner* (New York: Macmillan Publishing Company, 1982).

Page 28 Magic lantern picture after the set design by Joseph Hoffmann for the Bayreuth Festival of 1876. Historisches Museum, Frankfurt, Germany. From Richard Wagner, *Das Rheingold* libretto (Hamburg: Deutsche Grammophon Gmbh., 1989).

Page 32 Curtain design for the opening scene of the 1924 production of Wagner's *Das Rheingold*. From Richard Wagner, *Das Rheingold* libretto (London: EMI Records Ltd., 1989).

Pages 34-36 Illustrations © 1991 by Robin A. Smith.

Page 38 "Manius Curius Dentatus Refusing the Presents of the Samnite Ambassadors." Pen and black ink drawing, 1795-96, by Jean-Guillaume Moitte. From *The Metropolitan Museum of Art Bulletin*, Volume XLVIII, Number 2 (New York: The Metropolitan Museum of Art, 1990).

Page 41 Road from Eleusis to Athens. From Jim Harter, ed., *Images of World Architecture* (New York: Bonanza Books, 1990).

Page 45 Marketplace of Athens. Illustration courtesy of the Granger Collection.

Page 46 Illustration © 1991 by Sanie Holland.

Page 48 Thebes, silver stater, late 6th to early 5th century B.C. From Richard G. Doty, *Money of the World* (New York: Grosset & Dunlap, Inc. and The Ridge Press, Inc., 1978).

Page 49 Cowries, New Guinea. From A. Hingston Quiggin, *A Survey of Primitive Money: The Beginnings of Currency* (London: Methuen & Co. Ltd., 1949).

Page 52 "Amerika." Collage illustration © 1990 by Jim Harter. From Bryce Milligan, *Litany Sung at Hell's Gate* (San Antonio, Tex.: M & A Editions, 1990).

Page 55 "The Flower of the Wise." From the "Alchemical Manuscript" of 1550 in Basle University Library. From Titus Burckhardt, *Alchemy: Science of the Cosmos, Science of the Soul* (Rockbury, Mass. and Shaftesbury, England: Element Books, Ltd., 1986).

Page 60 From "The Spirit in the Bottle" reprinted from *The Complete Grimm's Fairy Tales* by permission of Pantheon Books, a division of Random House, Inc. By Jakob Ludwig Karl Grimm and Wilhelm Karl Grimm, illustrated by Josef Scharl. Copyright 1944 by Pantheon Books, Inc. and renewed 1972 by Random House, Inc.

Pages 64, 69 "Volunteers rebuilding the main *chorten* of Ganden Monastery," Northeast of Lhasa, 1987 (p. 64); "Loom in a nomad camp at Pekhu Tso," Western Tibet, 1988 (p. 69). Photographs by Galen Rowell. From His Holiness the Fourteenth Dalai Lama of Tibet, *My Tibet* (Berkeley: University of California Press, 1990).

Pages 70-75 Illustrations © 1991 by Sanie Holland.

Page 77 The planet Mercury as the god of commerce. From *The Golden Age of Dutch Manuscript Painting* (New York: George Braziller, Inc., 1990).

Page 78 Coin of Caesar. From Richard G. Doty, *Coins of the World* (New York: Grosset & Dunlap, Inc. and Ridge Press, Inc., 1978).

Page 80 Celtic coins. From Stephen Mitchell and Brian Reeds, eds., *Coins of England and the United Kingdom* (London: Seaby, 1990).

Page 82 "Coining." From Charles C. Gillispie, ed., *A Diderot Pictorial Encyclopedia of Trades and Industry* (New York: Dover Publications Inc., 1959).

Page 84 Illustration © 1991 by Robin A. Smith.

Page 87 18th-century watercolor illustration of a folio from "The Bhāgavata Purāna." From Pratapaditya Pal, *Nepal: Where the Gods are Young* (New York: The Asia Society, 1975).

Page 90 "Shiva and the Ganges." 18th-century Indian miniature. Alwar Museum, Alwar, Rajasthan. From *Aditi: The Living Arts of India* (Washington, D.C.: Smithsonian Institution Press, 1985).

Page 93 Illustration by William Schreiber. From *Walking into the Sun: Stories My Grandfather Told* (Oakland: California Health Publications, 1990).

Page 100 Coin with Tughra of Mustafā III, dated 1762. From Ariel Berman, *Islamic Coins* (Jerusalem: L.A. Mayer Memorial Institute for Islamic Art, 1976).

Page 144 "Sagittarius." Oxford, Bodleian Library, Ms. Marsh 144. From *Ars Orientalis: The Arts of Islam and the East, Vol. 3, 1959* (Washington, D.C. and Ann Arbor, Mich.: The Smithsonian Institution and University of Michigan).

PROFILES

DAVID APPELBAUM is a professor of philosophy at SUNY New Paltz. His most recent book, *Real Philosophy* (Penguin, 1991) was co-edited with Jacob Needleman.

SRI AUROBINDO (1872-1950), born in Calcutta and educated in England, became an anti-British political activist when he returned to India at the age of 21. During a time spent in jail for his activities, he became interested in yoga, and in 1910 he withdrew to Pondicherry where he devoted the rest of his life to developing his vision of spiritual evolution and the practice of Integral Yoga.

ROB BAKER is co-editor of PARABOLA Magazine.

THOMAS BUCKLEY is an associate professor of Anthropology at the University of Massachusetts, Boston. He has lived among the Native Americans of the Klamath River region of northwestern California off and on for many years, and is currently compiling a collection of his essays about them.

TITUS BURCKHARDT (1908-1984), German-Swiss author of the traditionalist school, served as artistic director of the Urs Graf Publishing House which published facsimiles of medieval Celtic manuscripts including The Book of Kells. His writings include *An Introduction to Sufi Doctrine*, *Sacred Art in East and West*, and *Fez, City of Islam*.

RENÉ GUÉNON (1886-1951), French scholar of religions, wrote extensively on Hinduism, Taoism, and Islam. His books include *The Crisis of the Modern World* and *Introduction to the Study of the Hindu Doctrines* (both Luzac).

JOSEPH KULIN is executive publisher of PARABOLA Magazine.

MOTHER TERESA, 1979 recipient of the Nobel Peace Prize, was born Agnes Gonxha Bojaxhiu of Albanian parents in Yugoslavia. In 1950 she founded the order of the Missionaries of Charity in Calcutta, and since then has established 450 houses to shelter the poor in ninety-five countries on five continents.

KARL H. POTTER is the editor of the *Encyclopedia of Indian Philosophies* (Princeton University Press), and *Presuppositions of India's Philosophies* (Prentice Hall).

E.F. SCHUMACHER (1911-1977) was a British economist and writer born in Germany. Founder and chairman of the Intermediate Technology Development Group, he wrote *Roots of Economics* (1962) and *A Guide for the Perplexed* (Harper & Row, 1978).

P.L. TRAVERS has been a consulting editor to PARABOLA since its inception. Her books include the Mary Poppins series, and she has also written extensively on myth and story. Her most recent collection, *What the Bee Knows* (Aquarian Press, 1990), includes many essays originally published in PARABOLA Magazine.

PHILIP ZALESKI is a freelance writer and senior editor of Parabola.

THE HUNTER

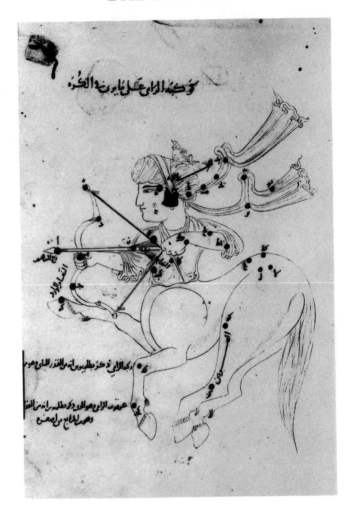